10 Alabama ACAP Grade 6 Math Practice Tests

The Ultimate Test Prep Collection with Answer Explanations

Dr. A. Nazari

10 Practice Tests

 The Grand Championship Collection 

Welcome, future Math Champion!

*You hold the **ultimate collection** —
ten full-length practice tests designed to take you
from first attempt to **complete mastery.***

🏅 *Conquer every Grade 6 topic*

🏅 *Build unshakeable confidence*

🏅 *Rise from Bronze to Gold to Champion*

🏅 *Arrive at test day fully prepared*

The championship begins now.

❝ *Ten tests may seem like
a marathon, but champi-
ons are made one step at a
time. Trust the process!* **❞**

 The Champion's Path

Your 4-phase journey from Bronze to Champion

Bronze Round (Tests 1–3)

*Your warm-up matches. Take these **untimed** to learn the format and set your baseline. Read the answer explanations after each test — this is where you build your foundation.*

Silver Round (Tests 4–6)

*Set a timer for **75 minutes**. Focus on the topics that tripped you up in Bronze. Practice showing your work on every problem. Your accuracy should be climbing.*

Gold Round (Tests 7–9)

*Full timed conditions (**60 minutes**). Simulate the real exam environment. Review only the questions you missed — targeted practice is the key to gold.*

Championship Final (Test 10)

Your final match. Full exam conditions — timed, quiet, no breaks. This is your victory lap. Show yourself how far you've come!

Your Championship Kit

- **10 Full-Length Practice Tests** — every Grade 6 topic
- **Formula Reference Sheet**
- **Complete Answer Key** with explanations
- **Championship Scoreboard** to track your rise

 **Champion's Tip:** *Space your tests 2–3 days apart. Use the days in between for targeted review. By Test 10, you'll be amazed at your transformation.*

The Champion's Playbook

Seven rules that separate champions from the rest

I   **Read every question twice.** *The first read tells you the topic. The second tells you exactly what to solve for. Champions never skim.*

II   **Mark the clues.** *Circle key numbers, underline the question, and cross out information that's just there to distract you.*

III   **Choose your strategy.** *Before touching pencil to paper, decide: Am I setting up a ratio? Solving an equation? Finding area? Name the approach.*

IV   **Solve, then match.** *For multiple choice — work the problem on scratch paper first, then find your answer among the choices.*

V   **Eliminate and conquer.** *Cross out obviously wrong answers. If you're left with two, you've already doubled your odds. Make an educated pick.*

VI   **Estimate to verify.** *After solving, ask: "Is this answer reasonable?" A quick mental estimate catches most calculation errors.*

VII   **Leave nothing blank.** *Even a well-reasoned guess is worth more than an empty space. Use partial work to support your answer.*

Timing Mastery

Tests 1–3: **Untimed** (build foundation)   *Tests 4–6:* **75 min** (build speed)   *Tests 7–10:* **60 min** (championship conditions)

⭐ Grade 6 Championship Topics

🏅 *Ratios & Proportions*   🏅 *Integers & Rational Numbers*   🏅 *Expressions & Equations*

🏅 *Geometry & Measurement*   🏅 *Statistics & Data Analysis*

 **❝** *A true champion isn't someone who never makes mistakes — it's someone who learns from **every single one**. After each test, review your errors carefully. That's where the real growth happens.* **❞**

Find more at
ViewMath.com/AL-Grade6

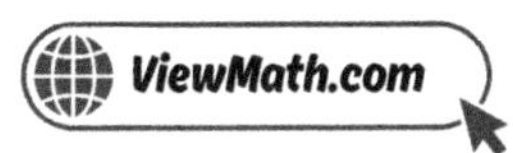

The Champion's Toolkit

Prepare your workspace before each championship round

Required Equipment

Sharpened Pencils — Two #2 pencils — champions always have a backup

Quality Eraser — A clean, soft eraser that won't smudge your work

Scratch Paper — Blank paper for calculations, diagrams, and number lines

Ruler — Essential for geometry and coordinate plane questions

Timer — Begin using from the Silver Round onward

Quiet Workspace — A calm, well-lit area free from distractions

Permitted in Competition

- ✓ Pencil and eraser
- ✓ Scratch paper (provided)
- ✓ Ruler (if specified)
- ✓ Formula reference in this book

Not Permitted

- ✗ Calculators
- ✗ Electronic devices
- ✗ Textbooks or notes
- ✗ Outside help

For Parents & Teachers

- With 10 tests, space them **2–3 days apart**. This gives time to review mistakes and study between rounds.

- Let your child take Tests 1–3 untimed to build familiarity and establish a baseline.

- After each test, go through the Answer Key together. Focus on **understanding the reasoning**, not memorizing answers.

- Use the Championship Scoreboard to visualize long-term progress. Celebrate improvements at every tier!

- Pair with our **Grade 6 Math Study Guide** for topics that need sustained attention.

Find more at
ViewMath.com/AL-Grade6

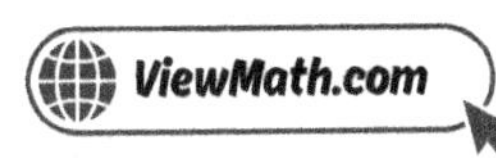

Formula Reference Sheet

🔺 Area Formulas

Rectangle $A = l \times w$

Parallelogram $A = b \times h$

Triangle $A = \dfrac{1}{2} \times b \times h$

Trapezoid $A = \dfrac{1}{2}(b_1 + b_2) \times h$

📦 Volume

Rectangular $V = l \times w \times h$

Prism

📦 Surface Area

Find the area of each face, then add them all up.

Rectangular Prism:

$SA = 2lw + 2lh + 2wh$

Order of Operations

P   Parentheses first

E   Exponents

M/D Multiply & Divide (left to right)

A/S Add & Subtract (left to right)

% Ratios & Percents

Ratio: $a : b$ or $\dfrac{a}{b}$

Unit rate: amount per 1 unit

Percent: a ratio out of 100

Part = Percent × Whole

⚖️ Integers & Absolute Value

Integers:

$\ldots, -3, -2, -1, 0, 1, 2, 3, \ldots$

$|-5| = 5 \quad |5| = 5$

Absolute value = distance from 0

X^1 Expressions & Equations

Exponent: $3^4 = 3 \times 3 \times 3 \times 3 = 81$

Variable: a letter that stands for a number

Equation: two expressions joined by $=$

Inequality: uses $<, >, \leq, \geq$

◎ Coordinate Plane

Ordered pair: (x, y)

x-axis: horizontal y-axis: vertical

Origin: $(0, 0)$

Four quadrants (I, II, III, IV)

📊 Statistics

Mean: sum of values ÷ count

Median: middle value (sorted)

Range: max − min

🏆 Championship Scoreboard 🏆

Track your rise through every championship round

Champion's Name: _______________________________

🏆 Round	🥇 Tier	📅 Date	⭐ Score	😊 Rating
1	Bronze			
2	Bronze			
3	Bronze			
4	Silver			
5	Silver			
6	Silver			
7	Gold			
8	Gold			
9	Gold			
10	👑			

X

My strongest topics (where I consistently score well):

____

____

Topics I improved on the most from Bronze to Gold:

____

____

My score trend (Bronze avg → Gold avg → Championship):

____

____

One strategy that helped me improve the most:

____

____

My confidence level for the real test (1–10): _________ / 10

Find more at
ViewMath.com/AL-Grade6

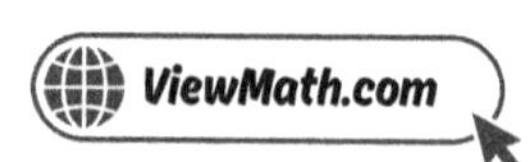

⭐ Table of Contents ⭐

Here's what we'll explore together!

Let's learn and have fun!

1

Practice Test 1

✅ *30 Questions*

✏️ Before You Start ✏️

- ✓ **Read each question carefully** before choosing your answer.
- ✓ **Show your work** on scratch paper when you need to.
- ✓ **Skip hard questions** and come back to them later.
- ✓ **Check your answers** when you're done.
- ✓ **Take your time** — there's no rush!

⭐ *You've Got This!* ⭐

Do your best and show what you know!

1. A garden has flowers and vegetables in a ratio of $4 : 1$. There are 20 flowers. How many vegetables are there?

2. The graph below shows the distance traveled by a delivery truck over time.

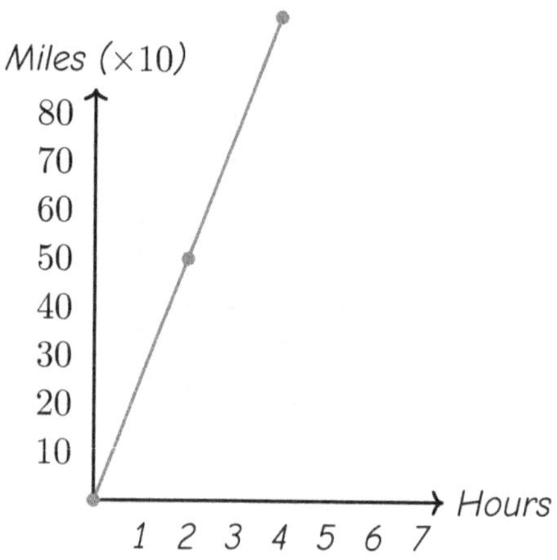

What is the truck's unit rate in miles per hour?

(A) 50 miles per hour

(B) 20 miles per hour

(C) 25 miles per hour

(D) 10 miles per hour

3. The double number line below shows equivalent ratios of scoops of mix to cups of water.

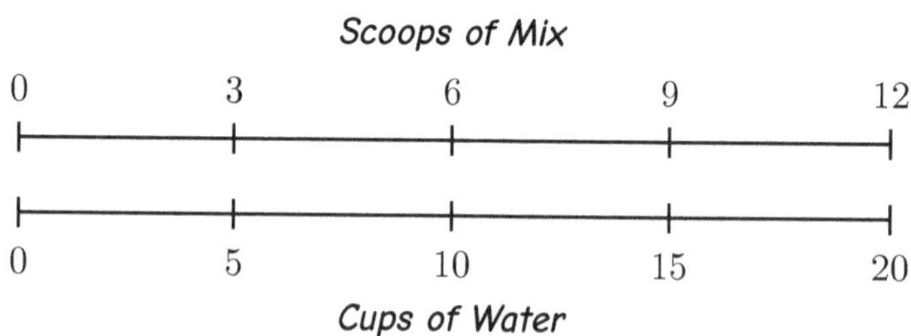

How many cups of water go with 9 scoops of mix?

(A) 10

(B) 12

(C) 15

(D) 18

Find more at
ViewMath.com/AL-Grade6

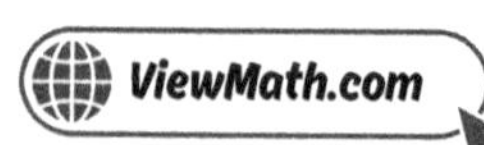

4. *A graph of a ratio relationship passes through $(0,0)$ and $(5,20)$. What is the unit rate?*

(A) 5

(B) 20

(C) 4

(D) 25

5. *The table shows the original prices and discount percents for three items.*

Item	Original Price	Discount
Backpack	$40	15%
Shoes	$60	10%
Hat	$20	25%

Which item has the greatest dollar amount of savings?

(A) *Backpack ($6.00)*

(B) *Shoes ($6.00)*

(C) *Hat ($5.00)*

(D) *Backpack and shoes are tied.*

6. *A bottle holds 750 mL. How many full bottles can you fill from a 6-liter jug?*

Your Answer:

7. *The number line below shows equal jumps from 0 to $\dfrac{5}{6}$.*

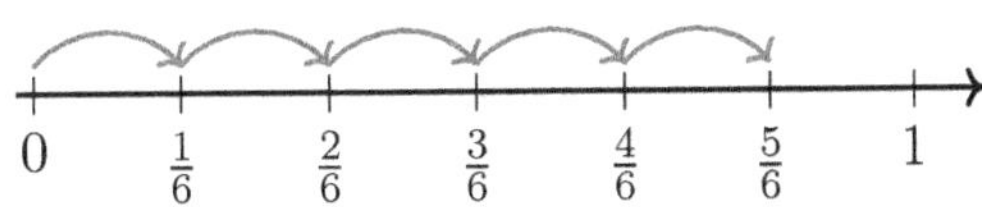

Which division problem does the number line represent?

(A) $\dfrac{5}{6} \div \dfrac{1}{6} = 5$

(B) $\dfrac{1}{6} \div \dfrac{5}{6} = 5$

(C) $\dfrac{5}{6} \times \dfrac{1}{6} = \dfrac{5}{36}$

(D) $\dfrac{5}{6} \div 5 = \dfrac{1}{6}$

8. Liam reads 1,344 pages in 8 weeks, reading the same number of pages each week. How many pages does he read per week?

  (A) 158   (B) 178

  (C) 186   (D) 168

9. A thermometer reads $0°C$. Is this temperature positive, negative, or neither? Explain.

Your Answer

10. Find $|-23|$.

11. Look at the input-output table below. Which expression was used to produce the output values?

Input (n)	Output
1	3
2	6
3	11
4	18

  (A) $n^2 + 2$ (B) $3n$

  (C) $n^2 + n + 1$ (D) $2n + 1$

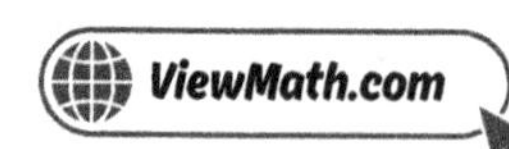

12. Look at the two-step diagram below. Write the expression that describes the final output when starting with the input n.

Your Answer

13. Write an expression that has 3 terms, a coefficient of 4, and a constant of 10.

Your Answer

14. Evaluate $3x + 7$ when $x = 4$.

(A) 19

(B) 34

(C) 15

(D) 47

15. Simplify: $3a + 7 + 5a - 2a + 3$

Your Answer

16. You have \$50 and spend d dollars. Which expression shows how much money you have left?

(A) $50 + d$

(B) $d - 50$

(C) $50d$

(D) $50 - d$

17. Write and solve an equation: A number decreased by 19 is 31.

Your Answer

Find more at
ViewMath.com/AL-Grade6

18. *Is $n = 4$ a solution to $n > 4$? Is $n = 4$ a solution to $n \geq 4$? Explain both.*

Your Answer

19. *Which value can you use to test that the graph of $x > -5$ is correct?*

(A) $x = -5$ *(it should make the inequality true)*   (B) $x = -6$ *(it should make the inequality true)*

(C) $x = 0$ *(it should make the inequality true)*   (D) $x = -10$ *(it should make the inequality true)*

20. *A triangular sail has a base of 3 m and a height of 7 m. How much fabric is needed to make 2 identical sails?*

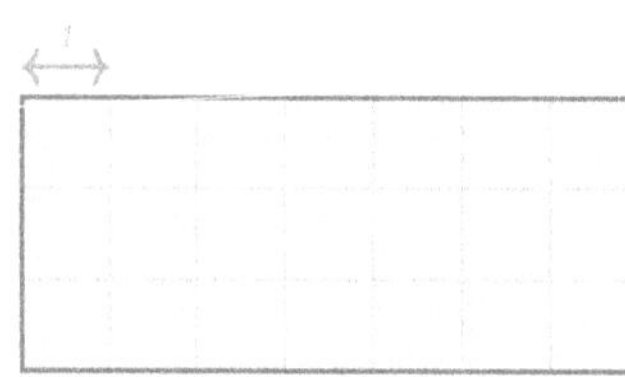

(A) $10.5 \ m^2$ (B) $42 \ m^2$

(C) $21 \ m^2$ (D) $20 \ m^2$

21. A parallelogram and a rectangle both have a base of 7 m and a height of 4 m. Which statement is true?

(A) The rectangle has a greater area.

(B) The parallelogram has a greater area.

(C) They have the same area.

(D) You cannot compare them.

22. Volume is measured in which type of units?

(A) Square units (cm^2)

(B) Linear units (cm)

(C) Cubic units (cm^3)

(D) No units are needed

23. A rectangle has vertices $(1, 2)$, $(7, 2)$, $(7, -4)$, and $(1, -4)$. What is the length of the longer side?

(A) 5 units

(B) 6 units

(C) 7 units

(D) 8 units

24. A composite figure is made of a 5×4 rectangle and a right triangle with base 5 and height 3, placed on top. What is the total area?

(A) 27.5 square units

(B) 35 square units

(C) 20 square units

(D) 32.5 square units

Find more at
ViewMath.com/AL-Grade6

25. A coach asks, "How many push-ups can each player do in one minute?" What type of data would this question produce?

   (A) Data that does not vary   (B) Data that varies from player to player

   (C) Non-numerical data only   (D) Exactly two data values

26. Look at the two data displays below.

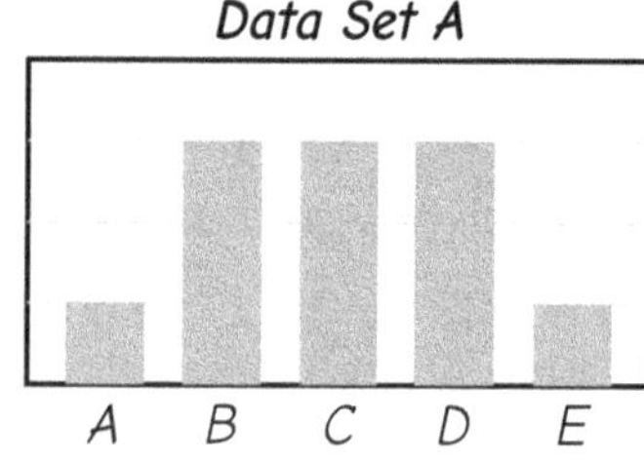

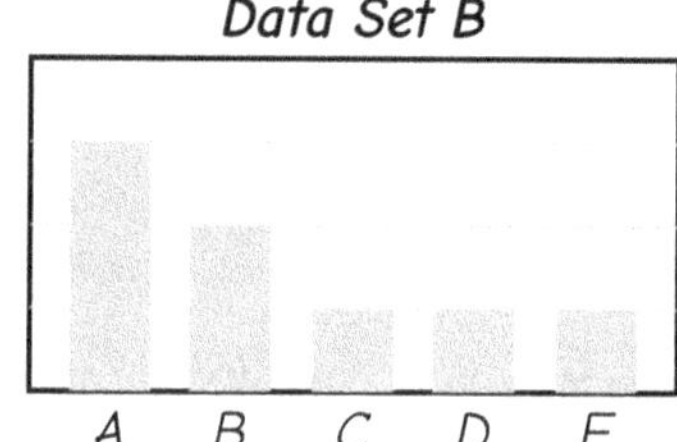

Which data set is skewed?

   (A) Data Set A is skewed right and Data Set B is symmetric.

   (B) Data Set A is symmetric and Data Set B is skewed right.

   (C) Both are symmetric.

   (D) Both are skewed.

27. A student says, "The mean and the median are always the same number." Is this correct?

   (A) Yes, they are always equal.

   (B) No, they are equal only when the data is perfectly symmetric.

   (C) No, the mean is always greater.

   (D) No, the median is always greater.

28. A histogram shows homework times: 0–14 min (5 students), 15–29 min (10 students), 30–44 min (8 students), 45–59 min (2 students). How many students are represented?

29. *Data:* 20, 25, 30, 35, 40, 45, 50, 55, 60. *What is Q3?*

(A) 40   (B) 45

(C) 50   (D) 52.5

30. *A student claims: "Our class did better because our range is smaller." Is range alone enough to support this claim? Explain.*

Your Answer:

 *End of Practice Test 1* 

Great job finishing the test!

🗒️ *My Score*

I got _________ out of 30 questions right.

*Check your answers in the **Answer Key** at the back of the book.*

💡 Review any questions you missed. That's how we learn!

📊 **Check Your Score Online!**

*Visit **ViewMath Academy** to enter your answers and see which topics you need to review. You can also explore lessons, take quizzes, track your scores, and save your progress!*

viewmath.com/score/6.1.AL.16

Or go to viewmath.com/score and enter code: 6.1.AL.16

2

Practice Test 2

30 Questions

✏ Before You Start ✏

- ✔ **Read each question carefully** before choosing your answer.
- ✔ **Show your work** on scratch paper when you need to.
- ✔ **Skip hard questions** and come back to them later.
- ✔ **Check your answers** when you're done.
- ✔ **Take your time** — there's no rush!

⭐ You've Got This! ⭐

Do your best and show what you know!

1. Look at the tape diagram below.

Boys: ▢▢▢▢
Girls: ▢▢▢▢▢▢

There are 24 girls. How many boys are there?

(A) 4

(B) 12

(C) 16

(D) 20

2. The table shows the earnings of two babysitters.

	Hours Worked	Earnings
Lily	5	$60
Noah	4	$52

Part A: Find each babysitter's unit rate (dollars per hour).

Part B: Who earns more per hour? How much more?

3. *The graph below shows points that represent equivalent ratios of flour to sugar in a recipe.*

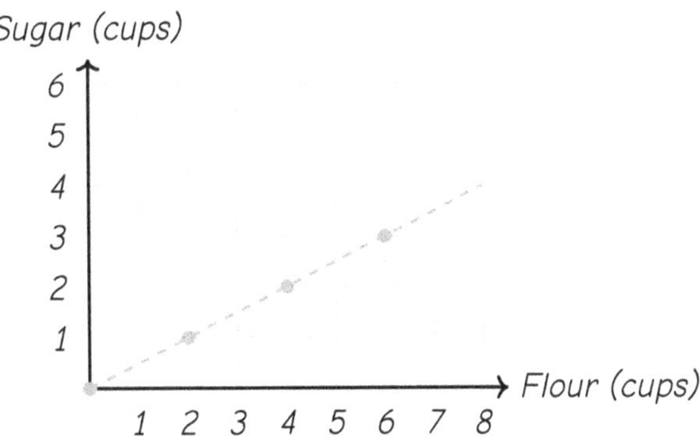

Part A: *What is the ratio of flour to sugar?*

Part B: *If you need 8 cups of flour, how many cups of sugar do you need?*

Your Answer:

4. *The points $(2, 5)$ and $(4, 10)$ are on a graph. What is the ratio $x : y$?*

  (A) $5 : 2$ (B) $1 : 2$

  (C) $2 : 5$ (D) $4 : 5$

5. *A town has 4,000 residents. 15% are children. How many children live in the town?*

  (A) 60   (B) 600

  (C) 150   (D) 1,500

6. *A race is 10 kilometers. How many centimeters is that?*

  (A) 100,000   (B) 10,000

  (C) 1,000,000   (D) 1,000

Find more at
ViewMath.com/AL-Grade6

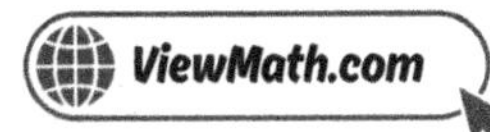

7. What is $\dfrac{3}{5} \div \dfrac{1}{5}$?

   (A) $\dfrac{3}{25}$ (B) 3

   (C) $\dfrac{1}{3}$ (D) 15

8. Compute $4{,}515 \div 15$. *Be careful with every digit of the quotient.*

   Your Answer:

9. Maria has $0 in her bank account. What does this mean?

   (A) She owes money.   (B) She has some savings.

   (C) She has no money and no debt.   (D) She has negative money.

10. A submarine is at -200 feet below sea level. How far is the submarine from sea level?

11. Evaluate: $3 \times (2 + 4)^2$

   (A) 42 (B) 108

   (C) 324 (D) 48

12. Write an expression for: "a number t divided by 4, then subtract 1."

   Your Answer:

13. A student made a factor tree for the expression $4(x+3)$, shown below. Fill in the missing pieces labeled A and B.

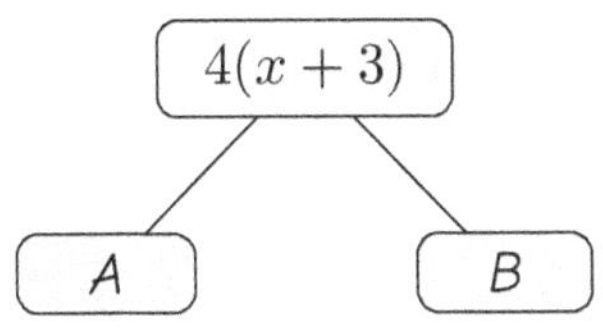

14. A parking garage charges \$5 plus \$3 per hour. How much does it cost to park for 6 hours? Use the expression $5 + 3h$.

15. Which expression is equivalent to $6a + 3 + 2a - 1$?

(A) $8a + 2$ (B) $8a + 4$

(C) $4a + 2$ (D) 10

16. In the formula $P = 4s$, the variable s represents the side length of a square. What does P represent?

(A) The area of the square   (B) The perimeter of the square

(C) The number of sides   (D) The diagonal of the square

17. Solve: $k + 15 = 32$

(A) $k = 47$ (B) $k = 17$

(C) $k = 27$ (D) $k = 2$

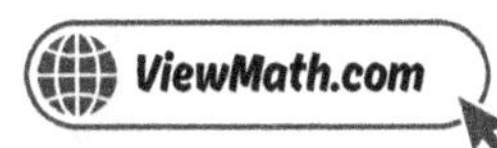

18. Which of the following is a solution to $n \leq 8$?

   (A) $n = 9$ (B) $n = 8.5$

   (C) $n = 8$ (D) $n = 10$

19. Write the inequality that matches this description: closed circle at -4, shade to the left.

20. A triangle has base 14 m and height 8 m. What is its area?

   (A) $112\ m^2$ (B) $22\ m^2$

   (C) $44\ m^2$ (D) $56\ m^2$

21. A parallelogram has base 8 m and height 3.5 m. What is the area?

   (A) $28\ m^2$ (B) $23\ m^2$

   (C) $14\ m^2$ (D) $11.5\ m^2$

22. A swimming pool is 25 m long, 10 m wide, and 2 m deep. How many cubic meters of water does it hold when full?

23. What is the distance between $(3, -1)$ and $(3, 7)$?

   (A) 6 units   (B) 7 units

   (C) 8 units   (D) 4 units

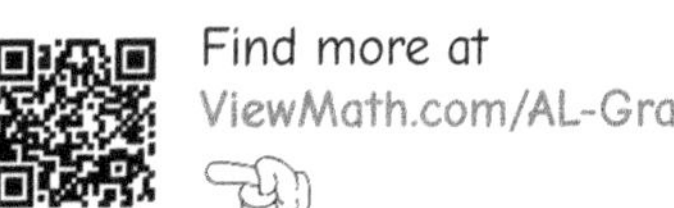

24. *A rectangle with vertices* $(1, 2)$, $(7, 2)$, $(7, 6)$, *and* $(1, 6)$ *has a right triangle cut from it with vertices* $(1, 2)$, $(7, 2)$, *and* $(7, 6)$. *What is the area of the remaining piece?*

Your Answer:

25. *A student surveyed classmates about how many books they read last month and made the dot plot below.*

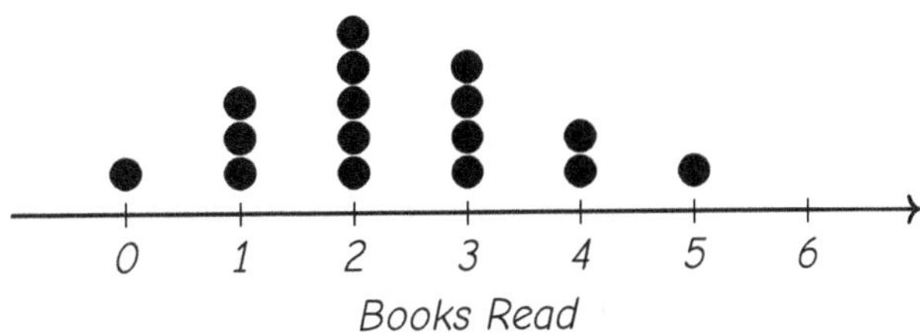

What does this dot plot confirm about the survey question "How many books did you read last month?"

(A) *It is not statistical because only whole numbers appear.*

(B) *It is statistical because the answers vary from 0 to 5.*

(C) *It is not statistical because some students read the same number.*

(D) *It is statistical because there are exactly 16 dots.*

26. *Look at the dot plot below showing students' quiz scores.*

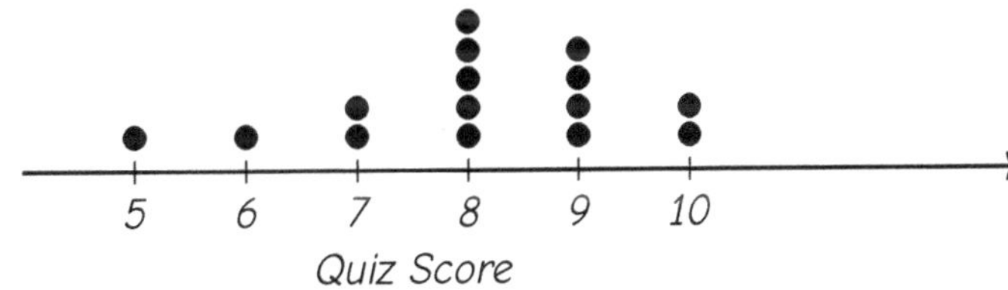

Describe the center, spread, and shape of the data. Identify where the data clusters and whether there are any gaps or outliers.

Your Answer:

27. What is the mean of the data set $4, 6, 8, 10, 12$?

(A) 6

(B) 8

(C) 10

(D) 12

28. A dot plot shows data clustered from 20 to 25 with no dots at 26, 27, 28, 29, and one dot at 30. What feature exists between 25 and 30?

(A) A cluster

(B) A peak

(C) A gap

(D) An interval

29. Look at the two box plots below showing test scores for two classes.

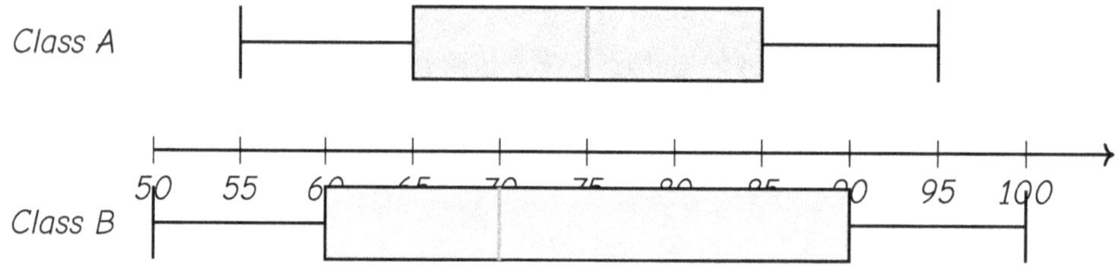

Which class has more consistent scores?

(A) Class A, because its box is narrower.

(B) Class B, because its box is wider.

(C) Both are equally consistent.

(D) Class B, because its median is higher.

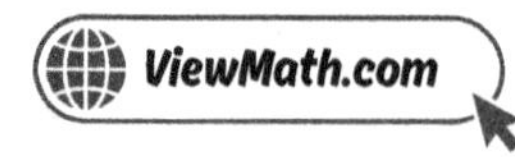

30. *The dot plots below show the number of sit-ups completed by students in two PE classes.*

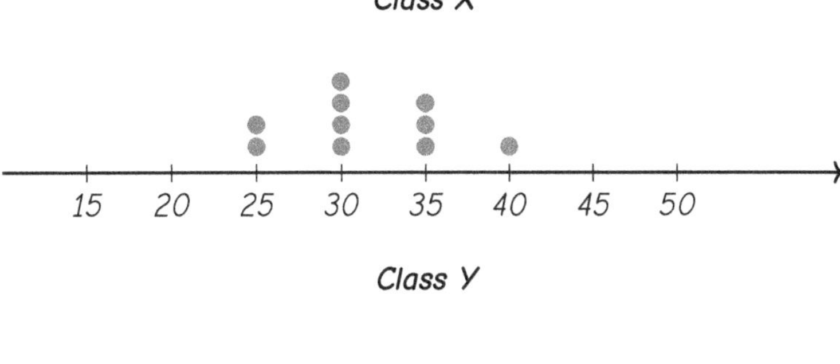

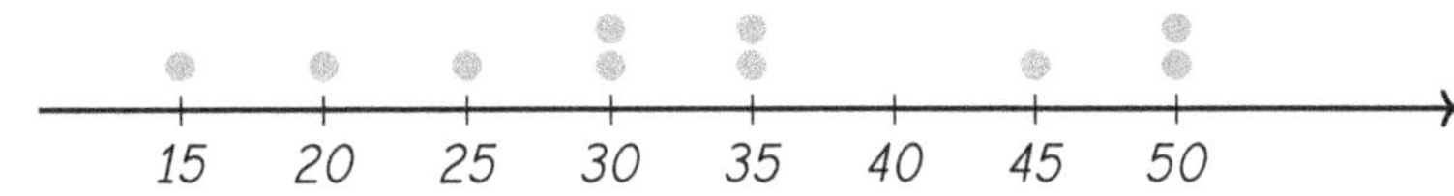

Which class has data that is more spread out?

(A) *Class X*

(B) *Class Y*

(C) *They have the same spread.*

(D) *Cannot be determined.*

 # End of Practice Test 2 

Great job finishing the test!

 **My Score**

I got _____________ out of 30 questions right.

Check your answers in the Answer Key at the back of the book.

💡 Review any questions you missed. That's how we learn!

📊 Check Your Score Online!

Visit **ViewMath Academy** to enter your answers and see which topics you need to review. You can also explore lessons, take quizzes, track your scores, and save your progress!

viewmath.com/score/6.1.AL.17

Or go to viewmath.com/score and enter code: 6.1.AL.17

3

Practice Test 3

30 Questions

✏️ Before You Start ✏️

- ✔ **Read each question carefully** before choosing your answer.
- ✔ **Show your work** on scratch paper when you need to.
- ✔ **Skip hard questions** and come back to them later.
- ✔ **Check your answers** when you're done.
- ✔ **Take your time** — there's no rush!

⭐ You've Got This! ⭐

Do your best and show what you know!

1. A baker uses 2 eggs **per** 3 cups of flour. Which ratio represents flour to eggs?

   (A) $2 : 3$ (B) $3 : 5$

   (C) $3 : 2$ (D) $2 : 5$

2. A box of 12 muffins costs $9. What is the cost per muffin?

   (A) $1.25   (B) $0.75

   (C) $1.33   (D) $3.00

3. A store sells 7 bananas for $2. How much do 21 bananas cost?

   (A) $4   (B) $6

   (C) $9   (D) $14

4. A ratio graph passes through $(6, 2)$. What is the value of y when $x = 15$?

   (A) 3   (B) 5

   (C) 7   (D) 10

5. What is 5% of 240?

   (A) 48   (B) 24

   (C) 12   (D) 120

6. Convert 3 miles to feet. (1 mile = 5,280 feet)

   (A) 5,280   (B) 10,560

   (C) 15,840   (D) 21,120

Find more at
ViewMath.com/AL-Grade6

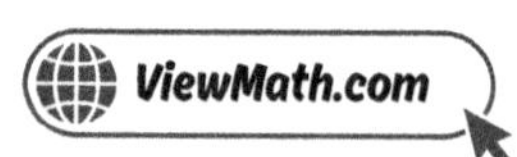

7. What is $\dfrac{1}{3} \div \dfrac{2}{3}$?

(A) $\dfrac{1}{2}$ (B) $\dfrac{2}{9}$

(C) $\dfrac{2}{3}$ (D) 2

8. What is $1{,}575 \div 5$?

(A) 305 (B) $3{,}150$

(C) 315 (D) 351

9. A helicopter is flying at 1,200 feet above sea level. Which integer best describes its altitude?

(A) $-1{,}200$ (B) 0

(C) $1{,}200$ (D) -12

10. Which expression gives the opposite of 5?

(A) $|5|$ (B) $5 + 5$

(C) $-(-5)$ (D) $-(5)$

11. Which shows $5 \times 5 \times 5$ written using an exponent?

(A) $5 + 3$ (B) 5×3

(C) 3^5 (D) 5^3

12. Write an expression for: "the sum of a and b, divided by 2."

Your Answer:

Find more at
ViewMath.com/AL-Grade6

13. How many terms are in the expression $7k$?

   (A) 1   (B) 2

   (C) 3   (D) 7

14. Evaluate $\dfrac{a-b}{5}$ when $a = 30$ and $b = 5$.

15. Use the distributive property to factor: $12x + 18$

   (A) $6(2x + 3)$ (B) $12(x + 18)$

   (C) $3(4x + 18)$ (D) $2(6x + 16)$

16. A plumber charges $50 for a house visit plus $40 per hour of work. Which of the following represents the cost of a 3-hour job?

   (A) $90   (B) $120

   (C) $170   (D) $190

Find more at
ViewMath.com/AL-Grade6

17. The table shows Mario's steps for solving an equation. In which step did he make an error?

Equation: $\dfrac{n}{5} = 8$

Step 1: Multiply both sides by 5

Step 2: $n = 8 + 5$

Step 3: $n = 13$

(A) No error — $n = 13$ is correct

(B) Step 1 — he should have divided both sides by 5

(C) Step 2 — multiplying by 5 gives $n = 8 \times 5 = 40$, not $8 + 5$

(D) Step 3 — $8 + 5 = 14$, not 13

18. A parking lot allows no more than 200 cars. Which inequality describes the number of cars c?

(A) $c > 200$

(B) $c \geq 200$

(C) $c < 200$

(D) $c \leq 200$

19. Which of the following graphs represents $x \leq -1$?

(A) Open circle at -1, shade right

(B) Closed circle at -1, shade right

(C) Open circle at -1, shade left

(D) Closed circle at -1, shade left

20. A triangular pennant has a base of 5 in and a height of 12 in. What is its area?

(A) $60\ in^2$

(B) $17\ in^2$

(C) $30\ in^2$

(D) $34\ in^2$

Find more at
ViewMath.com/AL-Grade6

21. *A trapezoid has bases 9 in and 15 in and height 6 in. What is the area?*

(A) $90\ in^2$ (B) $72\ in^2$

(C) $45\ in^2$ (D) $135\ in^2$

22. *An L-shaped solid is formed by joining two rectangular prisms as shown. What is the total volume?*

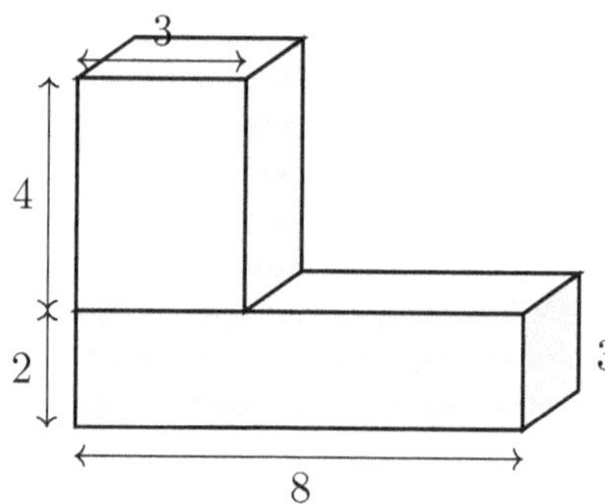

All depths are 3 units.

23. *A square has one vertex at $(-2, -2)$ and side length 5. The sides are horizontal and vertical. Which could be the opposite vertex?*

(A) $(3, 3)$ (B) $(3, -7)$

(C) $(5, 5)$ (D) $(2, 2)$

24. *A rectangle has vertices $(-5, -1)$, $(3, -1)$, $(3, 4)$, and $(-5, 4)$. Emma says the area is 30 square units, and Jake says it is 40 square units. Who is correct?*

(A) *Emma*   (B) *Jake*

(C) *Neither — the area is 35 square units.*   (D) *Neither — the area is 45 square units.*

Find more at
ViewMath.com/AL-Grade6

25. A librarian asks, "How many books were checked out today?" Is this a statistical question?

(A) Yes, because it is about books.

(B) Yes, if she plans to ask the same question over many days.

(C) No, because on one specific day there is only one answer.

(D) No, because libraries always have the same number of checkouts.

26. Describe the shape of the data set: $1, 2, 3, 3, 4, 4, 4, 5, 5, 5, 5$.

27. The mean of 6 quiz scores is 15. If one score of 9 is removed, what is the mean of the remaining 5 scores?

(A) 14.2

(B) 15

(C) 16.2

(D) 18

28. A dot plot shows quiz scores: 6 (2 dots), 7 (5 dots), 8 (4 dots), 9 (3 dots), 10 (1 dot). How many students took the quiz?

(A) 10

(B) 15

(C) 20

(D) 40

29. What percent of the data falls within the box of a box plot?

(A) 25%

(B) 50%

(C) 75%

(D) 100%

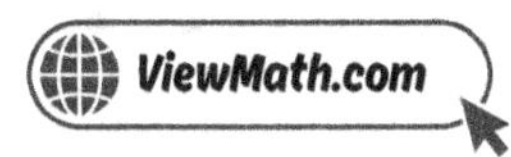

30. When two data sets have very little overlap, what can you say?

(A) The groups are very similar.

(B) The groups are clearly different from each other.

(C) Both groups have the same center.

(D) Both groups have the same spread.

Find more at
ViewMath.com/AL-Grade6

 # End of Practice Test 3 

Great job finishing the test!

 My Score

I got _____________ out of 30 questions right.

Check your answers in the **Answer Key** at the back of the book.

💡 Review any questions you missed. That's how we learn!

📊 Check Your Score Online!

Visit **ViewMath Academy** to enter your answers and see which topics you need to review. You can also explore lessons, take quizzes, track your scores, and save your progress!

viewmath.com/score/6.1.AL.18

Or go to viewmath.com/score and enter code: 6.1.AL.18

Practice Test 4

✅ 30 Questions

✏️ Before You Start ✏️

- ✓ **Read each question carefully** before choosing your answer.
- ✓ **Show your work** on scratch paper when you need to.
- ✓ **Skip hard questions** and come back to them later.
- ✓ **Check your answers** when you're done.
- ✓ **Take your time** — there's no rush!

⭐ You've Got This! ⭐

Do your best and show what you know!

1. A store sells 3 pencils **for every** 1 eraser. Which ratio represents pencils to erasers?

(A) $1:3$ (B) $3:1$

(C) $3:4$ (D) $1:4$

2. A bike travels at a unit rate of 14 miles per hour. How long will it take to travel 49 miles?

Your Answer

3. The ratio table below has an error in one row. Which row has the error?

x	y
4	10
8	20
12	25
16	40

(A) Row 1   (B) Row 2

(C) Row 3   (D) Row 4

4. On a graph, Line A passes through $(1,6)$ and Line B passes through $(1,4)$. Both go through the origin. Which line represents a faster rate? Explain.

Your Answer

5. What is 30% of 90?

(A) 3   (B) 27

(C) 60   (D) 30

Find more at
ViewMath.com/AL-Grade6

6. How many cups are in 3 quarts? $(1 \text{ quart} = 4 \text{ cups})$

   (A) 7   (B) 8

   (C) 10   (D) 12

7. If $\dfrac{a}{b} \div \dfrac{c}{d} = \dfrac{a}{b} \times \dfrac{?}{?}$, what fraction replaces the question marks?

   (A) $\dfrac{d}{c}$ (B) $\dfrac{c}{d}$

   (C) $\dfrac{b}{a}$ (D) $\dfrac{a}{b}$

8. A factory produces 5,376 bottles and packs them into cases of 32. How many cases are needed?

9. Which of the following numbers is negative?

   (A) 7   (B) 0

   (C) -4 (D) 12

10. What is $|-8|$?

    (A) -8 (B) 0

    (C) 8   (D) $-(-8)$

11. Evaluate: $20 - 4 \times 3 + 1$

    (A) 9   (B) 49

    (C) 51   (D) 7

Find more at
ViewMath.com/AL-Grade6

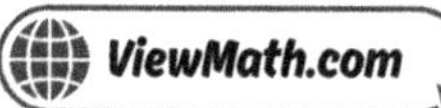

12. A notebook costs \$3. Write an expression for the cost of n notebooks plus a \$2 pen.

Your Answer

13. The table below shows different expressions. Which row has the coefficient identified incorrectly?

Row	Expression	Coefficient of the variable
1	$8p + 2$	8
2	$r - 5$	0
3	$3(x + 4)$	3
4	$10 + 6y$	6

(A) Row 1

(B) Row 2

(C) Row 3

(D) Row 4

14. Evaluate $m^2 + 2m + 1$ when $m = 4$.

Your Answer

15. Are $4(n + 2)$ and $4n + 8$ equivalent?

(A) Yes — they always give the same value

(B) No — they only match when $n = 2$

(C) No — $4(n + 2) = 4n + 2$

(D) Yes — but only for positive values of n

ViewMath.com

16. *The diagram shows a garden with a fixed fence on the left and new fencing needed for the other three sides. The garden is w feet wide and 12 feet long. Which expression gives the length of **new** fencing needed?*

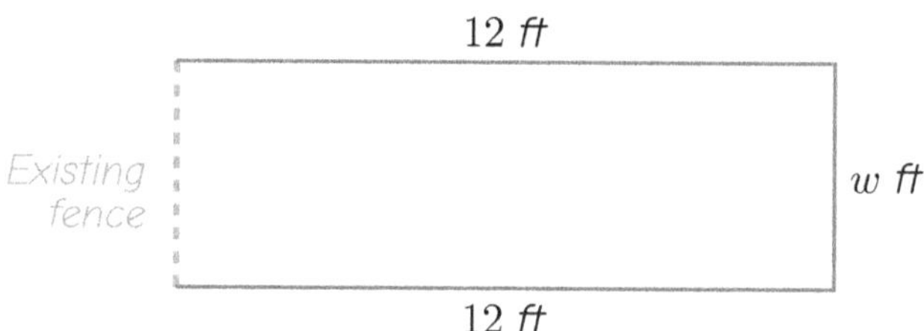

(A) $24 + w$

(B) $24 + 2w$

(C) $12 + 2w$

(D) $2(12 + w)$

17. *Solve:* $12p = 84$

(A) $p = 72$

(B) $p = 96$

(C) $p = 7$

(D) $p = 6$

18. *Which inequality represents "you need more than \$25 to buy the video game"?*

(A) $d \leq 25$

(B) $d < 25$

(C) $d > 25$

(D) $d \geq 25$

19. *Which inequality has a graph where $x = -4$ is in the shaded region but $x = -2$ is NOT?*

(A) $x > -3$

(B) $x < -3$

(C) $x \geq -3$

(D) $x \leq -3$

20. A triangle has base 20 in and height 7 in. What is the area?

(A) $140\ in^2$

(B) $27\ in^2$

(C) $70\ in^2$

(D) $54\ in^2$

21. A parallelogram has base 12 m and a slant side of 8 m. The height is 6 m. What is the area?

(A) $96\ m^2$

(B) $72\ m^2$

(C) $48\ m^2$

(D) $36\ m^2$

22. A rectangular prism has length $\frac{1}{2}$ m, width $\frac{1}{2}$ m, and height $\frac{1}{2}$ m. What is the volume?

(A) $\frac{1}{8}\ m^3$

(B) $\frac{1}{4}\ m^3$

(C) $\frac{3}{2}\ m^3$

(D) $\frac{1}{2}\ m^3$

23. Look at the triangle plotted below. What is the length of side $\overline{AB}$?

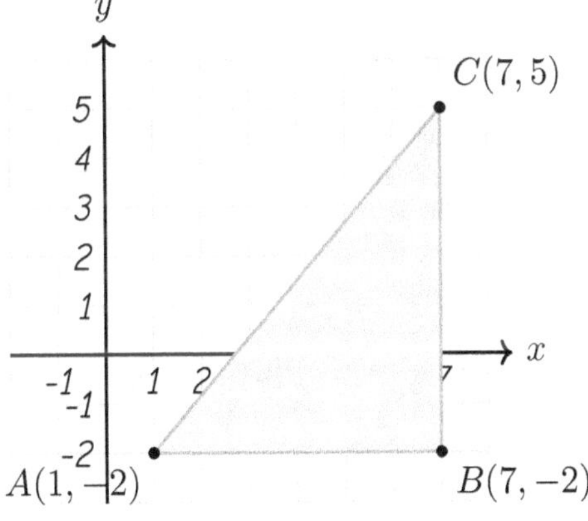

(A) 5 units

(B) 6 units

(C) 7 units

(D) 8 units

24. An L-shaped figure has vertices $(0,0)$, $(10,0)$, $(10,4)$, $(6,4)$, $(6,8)$, and $(0,8)$. What is the total area?

Your Answer

25. What makes a question statistical?

(A) It can be answered with a number.

(B) It is about a math topic.

(C) The answers are expected to vary.

(D) It has exactly one correct answer.

26. Two data sets have the same center but different spreads. Which statement is correct?

(A) The data sets must look the same on a graph.

(B) The data set with the larger spread has values farther from the center.

(C) Spread and center always change together.

(D) If the centers are the same, the spreads must also be the same.

27. Four students' heights in inches are: $58, 60, 62, 64$. A fifth student who is 82 inches tall joins. What happens to the mean?

(A) It stays at 61.

(B) It increases to 65.2.

(C) It decreases.

(D) It increases to 82.

28. A histogram has the following bars: 0–9 (height 3), 10–19 (height 7), 20–29 (height 10), 30–39 (height 5). How many data values are there in total?

(A) 4

(B) 10

(C) 25

(D) 39

29. The two box plots below represent daily sales (in dollars) at two stores over the same month.

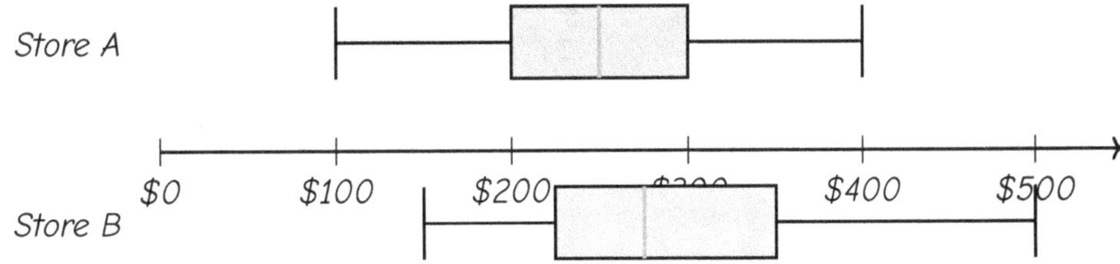

Compare the two stores. Which store has higher typical sales? Which has more predictable sales? Use the box plots to support your answers.

Your Answer

30. Runner A's race times (min): mean = 25, MAD = 1. Runner B's times: mean = 24, MAD = 5. Which runner is faster on average? Which is more consistent?

(A) Runner A is faster and more consistent.

(B) Runner B is faster but Runner A is more consistent.

(C) Runner B is faster and more consistent.

(D) They are the same speed.

 End of Practice Test 4 

Great job finishing the test!

☑ My Score

I got _____________ out of 30 questions right.

Check your answers in the Answer Key at the back of the book.

💡 *Review any questions you missed. That's how we learn!*

📊 Check Your Score Online!

Visit **ViewMath Academy** to enter your answers and see which topics you need to review. You can also explore lessons, take quizzes, track your scores, and save your progress!

viewmath.com/score/6.1.AL.19

Or go to viewmath.com/score and enter code: 6.1.AL.19

5

Practice Test 5

☑ 30 Questions

✏️ Before You Start ✏️

- ✔ **Read each question carefully** before choosing your answer.
- ✔ **Show your work** on scratch paper when you need to.
- ✔ **Skip hard questions** and come back to them later.
- ✔ **Check your answers** when you're done.
- ✔ **Take your time** — there's no rush!

⭐ You've Got This! ⭐

Do your best and show what you know!

1. A class votes on two activities: hiking and swimming. The ratio of votes for hiking to swimming is $3 : 2$. There are 25 votes total. How many voted for swimming?

Your Answer:

2. A car travels 240 miles using 8 gallons of gas. What is the unit rate in miles per gallon?

(A) 8

(B) 32

(C) 30

(D) 248

3. It takes 6 cups of flour to make 4 loaves of bread. How many cups of flour for 10 loaves?

(A) 12

(B) 10

(C) 15

(D) 20

Find more at
ViewMath.com/AL-Grade6

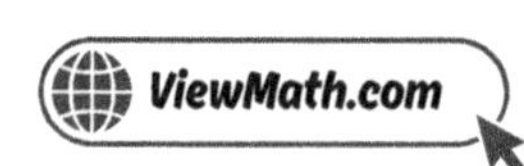

4. *The table and partial graph below show a ratio relationship.*

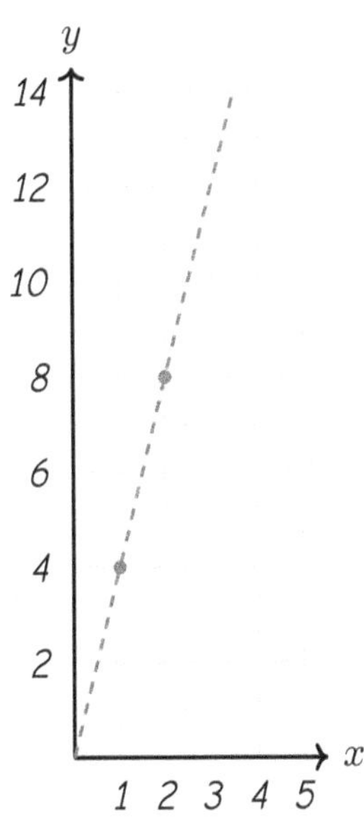

x	y
1	4
2	8
3	?

What is the missing value of y when $x = 3$?

(A) 10

(B) 12

(C) 14

(D) 16

5. *50% of what number is 33?*

(A) 16.5

(B) 66

(C) 83

(D) 99

6. *Convert 2 hours, 30 minutes to minutes, then to seconds.*

Your Answer:

Find more at
ViewMath.com/AL-Grade6

ViewMath.com

7. What is $\dfrac{2}{3} \div \dfrac{4}{5}$?

(A) $\dfrac{8}{15}$

(B) $\dfrac{6}{5}$

(C) $\dfrac{5}{6}$

(D) $\dfrac{2}{15}$

8. In the standard long-division algorithm, the four repeating steps in order are: Divide, _________, Subtract, Bring Down. What is the missing step?

(A) Estimate

(B) Multiply

(C) Check

(D) Add

9. Which real-world situation is best described by a positive number?

(A) A debt of \$50

(B) 20 feet below sea level

(C) A gain of 6 yards in football

(D) A temperature of 10° below zero

10. Name two different numbers whose absolute value is 11.

11. Evaluate: $7^2 - 10$

12. Which expression represents "the quotient of m and 5"?

(A) $5m$

(B) $m + 5$

(C) $m \div 5$

(D) $5 \div m$

Find more at
ViewMath.com/AL-Grade6

ViewMath.com

13. *In the expression $5(y + 8)$, name the two factors.*

14. *Evaluate w^3 when $w = 2$.*

(A) 6

(B) 8

(C) 16

(D) 32

15. *Simplify:* $5(2m + 3)$

(A) $7m + 3$

(B) $10m + 3$

(C) $10m + 15$

(D) $7m + 8$

16. *A phone battery starts at 100% and loses 8% each hour. Write an expression for the battery level after h hours.*

17. *Which inverse operation would you use to solve $\dfrac{n}{3} = 12$?*

(A) Divide both sides by 3

(B) Subtract 3 from both sides

(C) Add 3 to both sides

(D) Multiply both sides by 3

18. *Which phrase means $n \geq 12$?*

(A) n is less than 12

(B) n is greater than 12

(C) n is at least 12

(D) n is at most 12

Find more at
ViewMath.com/AL-Grade6

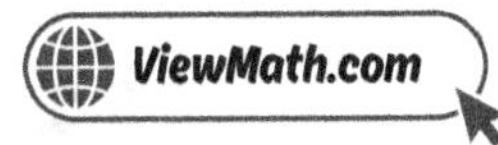

19. *Which of these numbers is a solution to $x \geq -3$?*

(A) -4 (B) -3.5

(C) -3 (D) -100

20. Two triangles both have a base of 10 cm. Triangle P has a height of 6 cm and Triangle Q has a height of 8 cm. How much greater is the area of Triangle Q?

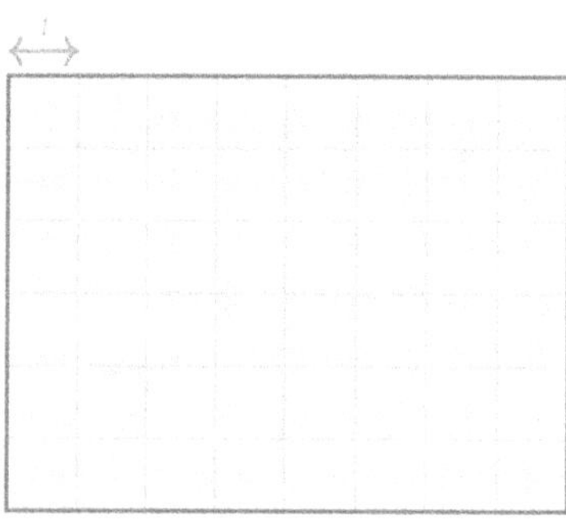

(A) $2\ cm^2$ (B) $10\ cm^2$

(C) $20\ cm^2$ (D) $40\ cm^2$

21. A trapezoid has an area of 84 cm^2 and bases of 10 cm and 14 cm. What is the height?

Your Answer

22. A rectangular prism has a volume of 120 cm^3. Its length is 10 cm and width is 4 cm. What is the height?

(A) 3 cm (B) 12 cm

(C) 6 cm (D) 30 cm

23. *A rectangle on the coordinate plane has a length of 10 units and width of 4 units. One vertex is at the origin $(0,0)$ and the sides are along the axes. Which of the following could be the opposite vertex?*

(A) $(10, 4)$

(B) $(14, 0)$

(C) $(4, 4)$

(D) $(10, 10)$

24. *A rectangle has vertices $(0,0)$, $(6,0)$, $(6,4)$, and $(0,4)$. What is the area?*

(A) 10 *square units*

(B) 20 *square units*

(C) 24 *square units*

(D) 12 *square units*

25. *A teacher asks, "How many pages did each student read last week?" This is a statistical question because —*

(A) *the answer is always a whole number*

(B) *every student reads the same amount*

(C) *the data collected would vary from student to student*

(D) *the question is about reading*

26. *Data: $15, 16, 16, 17, 17, 17, 18, 100$. How does the outlier 100 affect the center?*

(A) *It pulls the center toward 100, making it higher than typical values.*

(B) *It has no effect on the center.*

(C) *It makes the center equal to 100.*

(D) *It pulls the center lower.*

27. *Data: $18, 22, 25, 30$. Find the mean and median.*

Your Answer:

28. A dot plot has most of its dots between 40 and 50, with one dot at 10 and one at 80. What describes this data?

(A) Symmetric with no outliers

(B) Clustered near 40–50 with possible outliers at 10 and 80

(C) Uniform with equal spread

(D) Has a gap but no cluster

29. The data table below shows the number of books 9 students read this semester.

Student	1	2	3	4	5	6	7	8	9
Books	2	4	5	7	8	9	10	12	15

Find the five-number summary and the IQR.

30. Data: 10, 12, 14, 16, 18, 20, 22. The mean is 16. A student says the data is "very spread out." Is this correct?

(A) Yes, the range is 12.

(B) No, the range (12) is moderate and the values increase by a steady 2.

(C) Yes, because there are 7 values.

(D) No, because the mean equals the median.

Find more at
ViewMath.com/AL-Grade6

 ViewMath.com

 End of Practice Test 5 

Great job finishing the test!

My Score

I got _____________ out of 30 questions right.

Check your answers in the **Answer Key** at the back of the book.

Review any questions you missed. That's how we learn!

Check Your Score Online!

Visit **ViewMath Academy** to enter your answers and see which topics you need to review. You can also explore lessons, take quizzes, track your scores, and save your progress!

viewmath.com/score/6.1.AL.20

Or go to viewmath.com/score and enter code: 6.1.AL.20

Practice Test 6

30 Questions

✏ Before You Start ✏

- ✓ **Read each question carefully** before choosing your answer.
- ✓ **Show your work** on scratch paper when you need to.
- ✓ **Skip hard questions** and come back to them later.
- ✓ **Check your answers** when you're done.
- ✓ **Take your time** — there's no rush!

⭐ *You've Got This!* ⭐

1. A smoothie recipe uses 3 bananas **for every** 4 cups of yogurt. Write the ratio of bananas to yogurt. Then write the ratio of yogurt to bananas.

Your Answer

2. Brand X sells 5 notebooks for \$8.75. Brand Y sells 3 notebooks for \$4.50. Which is the better deal?

  (A) Brand X at \$1.75 each   (B) Brand Y at \$1.50 each

  (C) Brand X at \$1.50 each   (D) Brand Y at \$1.75 each

3. A teacher hands out 3 pencils for every 2 students. There are 16 students. How many pencils does she need?

  (A) 18   (B) 24

  (C) 20   (D) 32

4. A car travels at 30 miles per hour. Which set of points represents this ratio?

  (A) $(1, 30), (2, 60), (3, 90)$ (B) $(30, 1), (60, 2), (90, 3)$

  (C) $(1, 30), (2, 50), (3, 90)$ (D) $(1, 3), (2, 6), (3, 9)$

5. A store sells a book for \$25. Next week the price goes up by 10%. What is the new price?

  (A) \$35   (B) \$27.50

  (C) \$26   (D) \$22.50

6. Convert 7,000 grams to kilograms.

  (A) 0.7 kg   (B) 7 kg

  (C) 70 kg   (D) 700 kg

Find more at
ViewMath.com/AL-Grade6

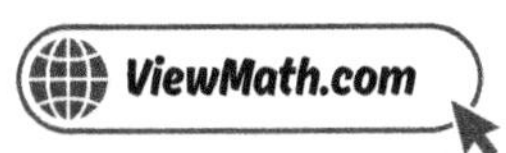

7. A recipe uses $\frac{2}{3}$ cup of sugar per batch. You have 4 cups of sugar. How many batches can you make?

  (A) $\frac{8}{3}$

  (B) 2

  (C) 6

  (D) 12

8. A student divides $3{,}612 \div 12$ and gets 31. The correct answer is 301. What mistake did the student most likely make?

  (A) The student forgot to subtract

  (B) The student skipped placing a zero in the quotient

  (C) The student used the wrong divisor

  (D) The student multiplied instead of dividing

9. The temperature outside is $-12°F$. What does this mean?

  (A) 12 degrees above zero

  (B) 12 degrees below zero

  (C) Exactly zero degrees

  (D) 12 degrees above freezing

10. What is $|-12|$?

  (A) -12

  (B) 12

  (C) 0

  (D) $-(-(-12))$

11. Evaluate: $2^3 + 3^2$

  (A) 13

  (B) 17

  (C) 25

  (D) 36

Get Online

Find more at
ViewMath.com/AL-Grade6

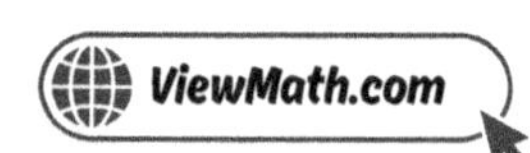

12. Look at the model below. Each box represents a value. Which expression matches the model?

$$\boxed{x}\;\boxed{x}\;\boxed{x}\;+\;\boxed{1}\;\boxed{1}$$

(A) $3x + 2$

(B) $2x + 3$

(C) $5x$

(D) $3 + 2x$

13. How many terms does the expression $4(a + b)$ have when written in its factored form?

14. Evaluate $4(a + b)$ when $a = 3$ and $b = 2$.

(A) 14

(B) 18

(C) 20

(D) 24

15. Which expression is equivalent to $9w - 4w + 6 - 6$?

(A) $5w$

(B) $5w + 12$

(C) $13w$

(D) 5

16. A pizza costs \$14 and is split equally among f friends. Which expression shows each friend's share?

(A) $14f$

(B) $14 + f$

(C) $14 - f$

(D) $14 \div f$

17. Solve: $6k = 54$. Then check your answer.

18. Name three solutions to the inequality $m \geq 6$.

Your Answer:

19. What is the difference between the graphs of $x > 5$ and $x \geq 5$?

(A) They shade in different directions

(B) $x > 5$ uses an open circle; $x \geq 5$ uses a closed circle

(C) $x > 5$ shades right; $x \geq 5$ shades left

(D) There is no difference

20. A triangle has base $\frac{3}{4}$ ft and height $\frac{2}{3}$ ft. What is its area?

21. A trapezoid has bases $b_1 = 4$ cm and $b_2 = 4$ cm with height 6 cm. What shape is this trapezoid actually equivalent to?

(A) A triangle

(B) A parallelogram

(C) A circle

(D) A pentagon

22. A rectangular prism has volume 180 cm^3, length 9 cm, and width 5 cm. What is the height?

23. Which pair of points forms a vertical segment?

(A) $(3,2)$ and $(7,2)$

(B) $(5,-1)$ and $(5,4)$

(C) $(1,3)$ and $(4,6)$

(D) $(0,0)$ and $(3,0)$

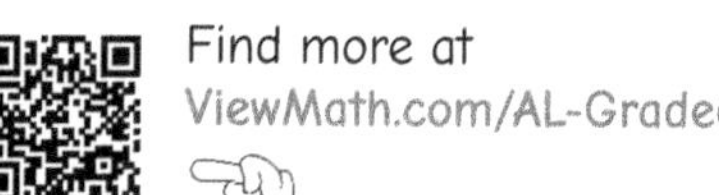
Find more at
ViewMath.com/AL-Grade6

24. To find the area of an irregular polygon on the coordinate plane, you can:

(A) Multiply all the coordinates together.

(B) Break it into rectangles and triangles, then add the areas.

(C) Count only the vertices.

(D) Subtract the perimeter from the largest coordinate.

25. Which question would produce data where every answer is the same?

(A) How many hours do you watch TV each day?

(B) How many feet are in one yard?

(C) What is your favorite sport?

(D) How far can you throw a ball?

26. Which data set has a gap?

(A) $10, 11, 12, 13, 14, 15$

(B) $2, 3, 4, 5, 6, 7$

(C) $20, 21, 22, 35, 36, 37$

(D) $8, 8, 8, 9, 9, 9$

27. The test scores of 4 students are $88, 92, 76, 84$. What is the mean?

(A) 84

(B) 85

(C) 86

(D) 88

28. A dot plot of daily temperatures shows: 70°F (1 dot), 72°F (3 dots), 74°F (5 dots), 76°F (4 dots), 78°F (2 dots). What is the range of the data?

(A) 4

(B) 6

(C) 8

(D) 15

29. *A box plot has min = 10, $Q1 = 20$, median = 30, $Q3 = 40$, max = 50. What is the IQR?*

(A) 10

(B) 20

(C) 30

(D) 40

30. *A good data summary should include —*

(A) *only the mean*

(B) *only the range*

(C) *both a measure of center and a measure of spread*

(D) *only the number of data values*

Find more at
ViewMath.com/AL-Grade6

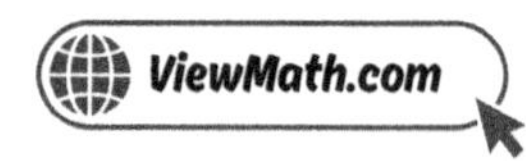

 End of Practice Test 6 

Great job finishing the test!

 My Score

I got _____________ out of 30 questions right.

*Check your answers in the **Answer Key** at the back of the book.*

Review any questions you missed. That's how we learn!

Check Your Score Online!

*Visit **ViewMath Academy** to enter your answers and see which topics you need to review. You can also explore lessons, take quizzes, track your scores, and save your progress!*

viewmath.com/score/6.1.AL.21

Or go to viewmath.com/score and enter code: 6.1.AL.21

Practice Test 7

 30 Questions

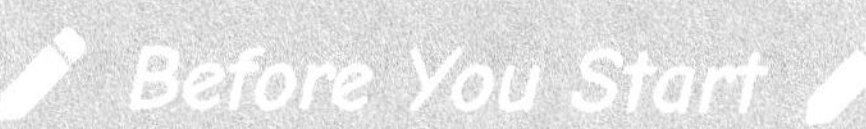

✏️ Before You Start ✏️

- ✓ **Read each question carefully** before choosing your answer.
- ✓ **Show your work** on scratch paper when you need to.
- ✓ **Skip hard questions** and come back to them later.
- ✓ **Check your answers** when you're done.
- ✓ **Take your time** — there's no rush!

⭐ You've Got This! ⭐

Do your best and show what you know!

1. A lemonade stand sells 7 cups of lemonade **for each** 2 cups of iced tea. If they sold 21 cups of lemonade, how many cups of iced tea did they sell?

Your Answer:

2. A garden hose fills a 150-gallon tank in 5 hours. What is the unit rate?

(A) 25 gallons per hour (B) 30 gallons per hour

(C) 35 gallons per hour (D) 750 gallons per hour

3. Complete the ratio table for the ratio 5 : 8.

5	8
10	?
?	32

Your Answer:

4. A line passes through the origin and the point $(5, 15)$. What is the ratio x to y?

(A) 1 : 5 (B) 5 : 1

(C) 1 : 3 (D) 3 : 1

5. A video game costs $60. Sales tax is 8%. What is the total cost?

Your Answer:

6. How many inches are in 5 feet?

(A) 50

(B) 55

(C) 60

(D) 72

7. A carpenter has $2\frac{1}{2}$ feet of wood. Each shelf needs $\frac{5}{8}$ of a foot. How many shelves can the carpenter cut?

8. What is $4{,}752 \div 12$?

(A) 396

(B) 369

(C) 406

(D) 39 R6

9. A submarine is 400 feet below sea level. Which integer represents the submarine's position?

(A) 400

(B) -400

(C) 0

(D) -40

10. Which number has an absolute value of 6?

(A) 6 only

(B) -6 only

(C) Both 6 and -6

(D) No number has an absolute value of 6

11. Evaluate: $5^2 + 4^2$

Find more at
ViewMath.com/AL-Grade6

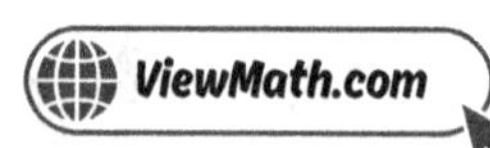

12. Sam has x stickers. He gives away 6. Which expression shows how many stickers Sam has now?

(A) $x + 6$

(B) $6x$

(C) $6 - x$

(D) $x - 6$

13. In the expression $2x + 5y + 10$, which of the following is true?

(A) The coefficient of x is 5

(B) There are 2 terms

(C) The constant is 10

(D) The coefficient of y is 10

14. A taxi charges \$3 plus \$2 per mile. The expression $3 + 2m$ gives the total fare. What is the fare for a 7-mile ride?

(A) \$12

(B) \$14

(C) \$17

(D) \$21

15. Simplify: $9x + 2 + 3x - 7$

Your Answer:

16. A rectangle has length l and width 5. Which expression represents its area?

(A) $l + 5$

(B) $2l + 10$

(C) $5l$

(D) $l - 5$

17. Solve: $x + 9 = 15$

(A) $x = 24$

(B) $x = 6$

(C) $x = 4$

(D) $x = 9$

Find more at
ViewMath.com/AL-Grade6

18. *Which phrase matches the inequality $t \leq 30$?*

   (A) *The temperature is more than 30 degrees*   (B) *The temperature is exactly 30 degrees*

   (C) *The temperature is no more than 30 degrees*   (D) *The temperature is at least 30 degrees*

19. *A graph shows a closed circle at -1 and shading to the left. Which number is NOT a solution?*

   (A) -1 (B) -5

   (C) -10 (D) 0

20. *A triangular flower bed has a base of 4.5 m and a height of 6 m. What is its area?*

21. *A park is shaped like a trapezoid with parallel sides of 40 m and 60 m and a height of 30 m. What is the area of the park?*

   (A) $1,800 \ m^2$ (B) $1,200 \ m^2$

   (C) $2,400 \ m^2$ (D) $1,500 \ m^2$

22. *A fish tank is 20 in long, 10 in wide, and 12 in tall. What is the volume of the tank?*

   (A) $42 \ in^3$ (B) $240 \ in^3$

   (C) $1,200 \ in^3$ (D) $2,400 \ in^3$

23. *What is the distance between $(-7, 3)$ and $(5, 3)$?*

Find more at
ViewMath.com/AL-Grade6

24. What is the area of the shaded right triangle on the coordinate plane?

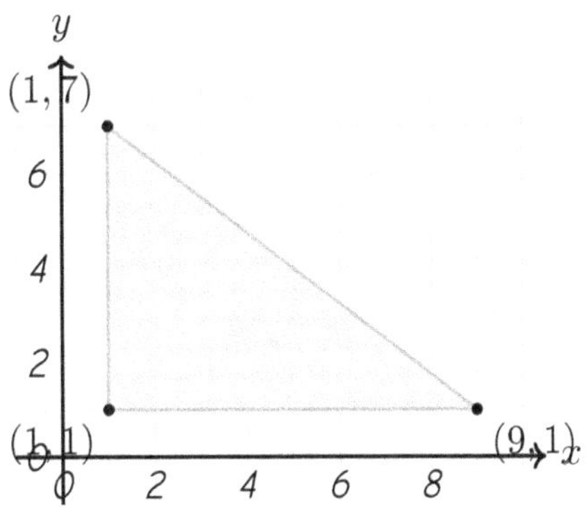

(A) 48 square units

(B) 24 square units

(C) 36 square units

(D) 12 square units

25. Give an example of a non-statistical question about sports.

Your Answer

26. Data set A: $20, 21, 22, 23, 24$. Data set B: $10, 15, 22, 29, 34$. Both have a center near 22. What is different?

(A) Data set A is more spread out.

(B) Data set B is more spread out.

(C) Both have the same spread.

(D) Neither data set has a center.

27. The table below shows the number of points scored by a basketball player in 8 games.

Game	1	2	3	4	5	6	7	8
Points	12	18	15	22	14	16	20	35

Find the mean and median. Which measure better represents the player's typical performance? Explain.

Your Answer

Find more at
ViewMath.com/AL-Grade6

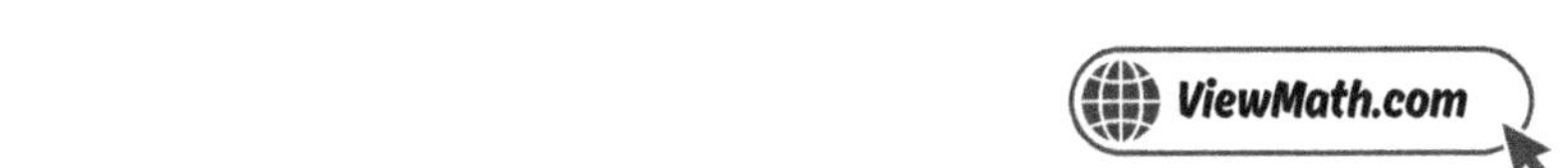

28. Which type of data display shows every individual data value?

(A) Histogram

(B) Circle graph

(C) Dot plot

(D) Box plot

29. If the whiskers of a box plot are both short and the box is narrow, what does this tell you?

(A) The data is very spread out.

(B) The data is highly consistent with little variability.

(C) There are many outliers.

(D) The data set is very large.

30. Group P: median $= 45$, IQR $= 20$. Group Q: median $= 60$, IQR $= 5$. Which group performed higher and which is more consistent?

(A) Group P is higher and more consistent.

(B) Group Q is higher and more consistent.

(C) Group P is higher, Group Q is more consistent.

(D) Group Q is higher, Group P is more consistent.

Find more at
ViewMath.com/AL-Grade6

⭐ End of Practice Test 7 ⭐

Great job finishing the test!

 My Score

I got __________ out of 30 questions right.

Check your answers in the Answer Key at the back of the book.

💡 *Review any questions you missed. That's how we learn!*

📊 **Check Your Score Online!**

Visit **ViewMath Academy** to enter your answers and see which topics you need to review. You can also explore lessons, take quizzes, track your scores, and save your progress!

viewmath.com/score/6.1.AL.22

Or go to viewmath.com/score and enter code: 6.1.AL.22

Practice Test 8

30 Questions

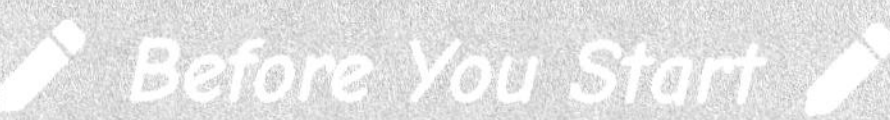

✏️ Before You Start ✏️

- ✓ **Read each question carefully** before choosing your answer.
- ✓ **Show your work** on scratch paper when you need to.
- ✓ **Skip hard questions** and come back to them later.
- ✓ **Check your answers** when you're done.
- ✓ **Take your time** — there's no rush!

⭐ You've Got This! ⭐

1. The ratio of cats to dogs at a pet store is $3 : 4$. Each part in the tape diagram represents 2 animals. How many dogs are there?

(A) 6

(B) 8

(C) 3

(D) 4

2. The graph shows the number of laps a swimmer completes over time.

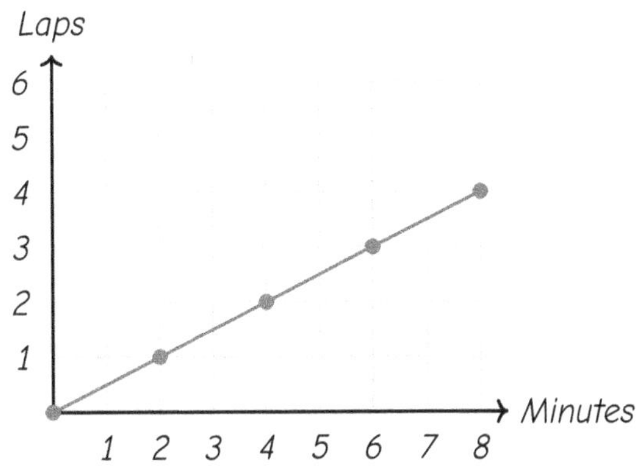

Part A: What is the swimmer's unit rate in laps per minute?

Part B: How many laps will the swimmer complete in 14 minutes?

Your Answer:

3. A smoothie recipe uses 2 cups of strawberries for every 3 cups of yogurt. How many cups of yogurt are needed for 10 cups of strawberries?

Your Answer:

4. A ratio graph passes through $(0, 0)$ and $(4, 7)$. What is y when $x = 8$?

(A) 11

(B) 14

(C) 15

(D) 12

5. *A school has 500 students. 72% passed the state exam. How many students passed?*

6. *The bar model below compares different length units.*

1 yard		
1 ft	1 ft	1 ft

If a rope is 7 yards long, how many feet is it?

(A) 10

(B) 14

(C) 21

(D) 28

7. *Kai says $\dfrac{2}{7} \div \dfrac{5}{6} = \dfrac{12}{35}$. Is Kai correct?*

(A) *No, the answer is $\dfrac{10}{42}$*

(B) *Yes, $\dfrac{12}{35}$ is correct*

(C) *No, the answer is $\dfrac{35}{12}$*

(D) *No, the answer is $\dfrac{7}{15}$*

8. *What is $7{,}344 \div 24$?*

(A) 36

(B) 360

(C) 306

(D) 3,006

9. *A scuba diver is 75 feet below sea level. Which integer represents the diver's depth?*

(A) 75

(B) −75

(C) +75

(D) 0

Find more at
ViewMath.com/AL-Grade6

ViewMath.com

10. Is $|-3|$ greater than, less than, or equal to $|3|$? Explain your reasoning.

11. The table below shows powers of 3. What value belongs in the blank?

3^1	3^2	3^3	3^4	3^5
3	9	27	?	243

(A) 36

(B) 64

(C) 81

(D) 108

12. Which expression represents "the sum of 9 and the product of 4 and t"?

(A) $9 + 4 + t$

(B) $9 + 4t$

(C) $9(4 + t)$

(D) $9 \times 4t$

13. In the expression $12k + 6$, what are the factors of the term $12k$?

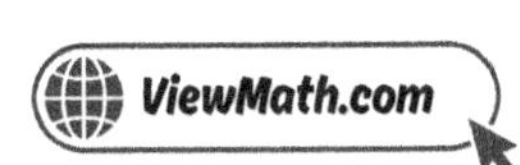

14. *The triangle below has a base of $b = 10$ cm and a height of $h = 6$ cm. Use the formula $A = \dfrac{1}{2} \times b \times h$ to find its area.*

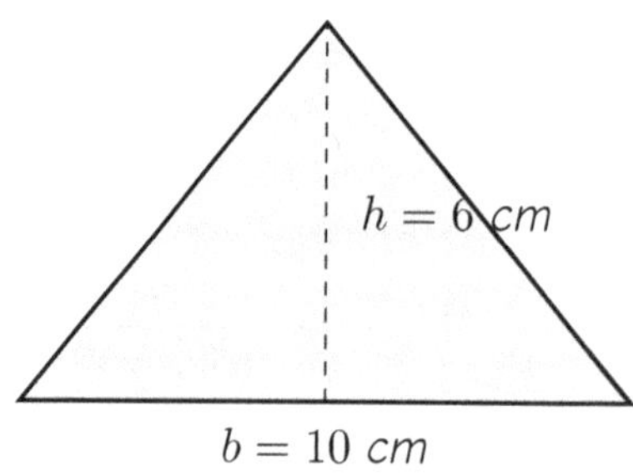

15. *Simplify:* $2(4m + 1) + 3m$

16. *A concert hall has r rows with 20 seats in each row plus 15 VIP seats in the front. Write an expression for the total number of seats.*

17. *Solve:* $\dfrac{w}{8} = 9$

18. *Which symbol correctly completes the statement? "The speed limit is 65 mph. You must drive _____ 65 mph."*

(A) $>$

(B) $<$

(C) $\leq$

(D) $\geq$

Find more at
ViewMath.com/AL-Grade6

ViewMath.com

19. A student says the graph of $x < 8$ and the graph of $x \leq 8$ look exactly the same. Is this correct?

Your Answer:

20. Look at the two triangles on the grid. Which triangle has a greater area?

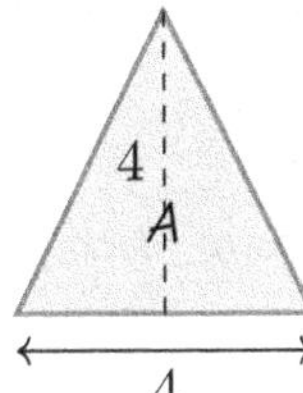
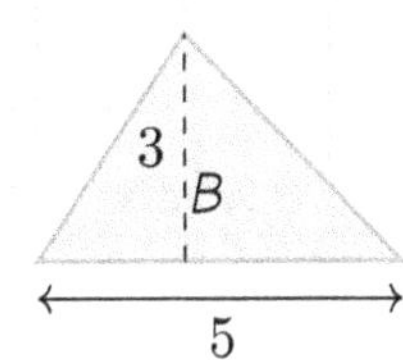

(A) Triangle A

(B) Triangle B

(C) They have the same area.

(D) Not enough information to tell.

21. A parallelogram has base 4.5 m and height 6 m. What is the area?

Your Answer:

22. Which formula gives the volume of a rectangular prism?

(A) $V = l + w + h$

(B) $V = 2lw + 2lh + 2wh$

(C) $V = l \times w \times h$

(D) $V = l \times w$

23. Two points share the same y-coordinate. They form a:

(A) Vertical segment

(B) Diagonal segment

(C) Horizontal segment

(D) There is not enough information to tell.

Find more at
ViewMath.com/AL-Grade6

ViewMath.com

24. A rectangle has vertices $(2, -1)$, $(2, 5)$, $(9, 5)$, and $(9, -1)$. What is the area?

Your Answer:

25. Explain why "What is my teacher's age?" is **not** a statistical question.

Your Answer:

26. Test scores: $68, 70, 72, 73, 74, 75, 76, 98$. Where is the center of this data?

(A) Around 68–70

(B) Around 73–75

(C) Around 83

(D) Around 98

27. Data: $3, 4, 5, 6, 7, 8, 60$. Find the mean and median. Which better describes a typical value?

Your Answer:

28. Which display would you use to show the heights of 200 students?

(A) Dot plot, because it shows individual values.

(B) Histogram, because it groups the large data set into intervals.

(C) Either works equally well.

(D) Neither works for numerical data.

29. A box plot shows: $min = 25$, $Q1 = 35$, $median = 50$, $Q3 = 65$, $max = 80$. Find the range and IQR.

Your Answer:

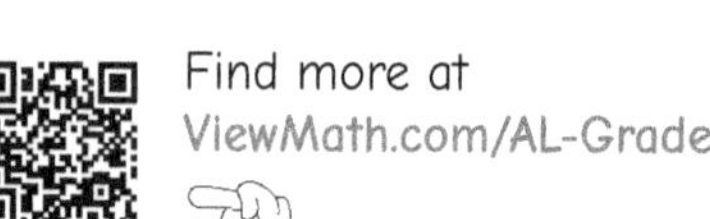
Find more at
ViewMath.com/AL-Grade6

30. *Two classes both have a mean of 75. Class A: MAD = 3. Class B: MAD = 9. A student from each class is picked at random. Whose score is more likely to be close to 75?*

(A) *The student from Class A*

(B) *The student from Class B*

(C) *Both equally likely*

(D) *Cannot determine*

Find more at
ViewMath.com/AL-Grade6

 End of Practice Test 8 

Great job finishing the test!

☑ My Score

I got _____________ out of 30 questions right.

Check your answers in the Answer Key at the back of the book.

💡 *Review any questions you missed. That's how we learn!*

📊 Check Your Score Online!

Visit **ViewMath Academy** to enter your answers and see which topics you need to review. You can also explore lessons, take quizzes, track your scores, and save your progress!

viewmath.com/score/6.1.AL.23

Or go to viewmath.com/score and enter code: 6.1.AL.23

9

Practice Test 9

📋 30 Questions

✏️ Before You Start ✏️

- ✓ **Read each question carefully** before choosing your answer.
- ✓ **Show your work** on scratch paper when you need to.
- ✓ **Skip hard questions** and come back to them later.
- ✓ **Check your answers** when you're done.
- ✓ **Take your time** — there's no rush!

⭐ You've Got This! ⭐

Do your best and show what you know!

1. The ratio of red to blue to green beads is $1 : 3 : 2$. There are 18 beads total. How many blue beads are there?

   (A) 3

   (B) 6

   (C) 9

   (D) 12

2. A printer prints 120 pages in 4 minutes. What is the unit rate?

   (A) 4 pages per minute

   (B) 30 pages per minute

   (C) 120 pages per minute

   (D) 480 pages per minute

3. Look at this ratio table. What is the missing value?

Apples	Oranges
3	5
6	?

   (A) 8

   (B) 10

   (C) 15

   (D) 11

4. Does the point $(3, 9)$ belong on the graph of the ratio $1 : 3$?

   (A) No, because $3 + 9 \neq 1 + 3$.

   (B) Yes, because $1 : 3 = 3 : 9$.

   (C) No, because $3 \times 3 \neq 1$.

   (D) Yes, because $3 + 9 = 12$.

5. A bike normally costs \$150. It is on sale for 20% off. What is the sale price?

   (A) \$30

   (B) \$120

   (C) \$130

   (D) \$100

Find more at
ViewMath.com/AL-Grade6

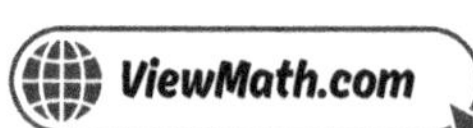

6. A recipe needs 2.5 liters of water. How many milliliters is that?

(A) 25   (B) 250

(C) 2,500   (D) 25,000

7. Each serving of juice is $\frac{2}{5}$ cup. A jug holds $\frac{4}{5}$ cup. How many servings are in the jug?

(A) $\frac{8}{25}$ (B) $\frac{2}{5}$

(C) $\frac{5}{2}$ (D) 2

8. The long division below for $1{,}575 \div 5$ is partially completed. A digit in the quotient is missing.

$$
\begin{array}{r}
3\ \square\ 5 \\
5\,\overline{\smash{\big)}\,1\,5\,7\,5} \\
\underline{-1\,5} \\
7 \\
\underline{-5} \\
2\,5 \\
\underline{-2\,5} \\
0
\end{array}
$$

What digit belongs in the box? Explain how you know.

Your Answer:

9. Which situation is best represented by the integer -15?

(A) A deposit of \$15 into a bank account   (B) A temperature of 15°F above zero

(C) An elevation of 15 feet above sea level   (D) A withdrawal of \$15 from a bank account

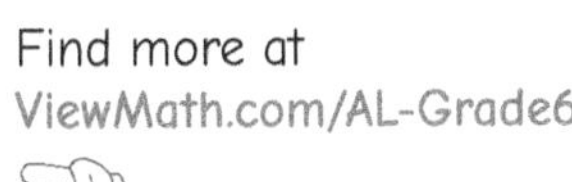
Find more at
ViewMath.com/AL-Grade6

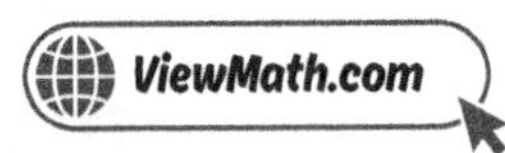

10. Three points are plotted on the number line below: A, B, and C.

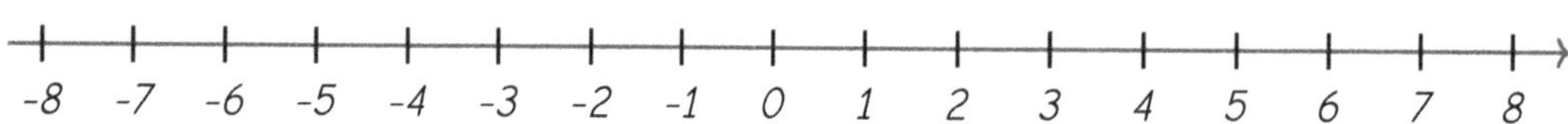

[A, above]-7 [B, above]3 [C, above]-4

Part A: Find the absolute value of each point: $|A|$, $|B|$, and $|C|$.

Part B: Which point is farthest from zero?

Part C: Which point is closest to zero?

11. Evaluate: $(5 - 1)^2 + 7$

(A) 14

(B) 23

(C) 11

(D) 33

12. Which expression represents "a number n plus 8"?

(A) $n - 8$

(B) $8n$

(C) $n + 8$

(D) $n \div 8$

13. Which statement about the expression $12 + 4w - 3$ is true?

(A) 12 is a coefficient

(B) w is a constant

(C) 4 is the coefficient of w

(D) There are 4 terms

14. Evaluate $n^2 - 5$ when $n = 6$.

(A) 7

(B) 31

(C) 41

(D) 25

Find more at
ViewMath.com/AL-Grade6

ViewMath.com

15. *Which expressions are equivalent? Test with* $x = 2.$
Expression A: $3(x + 1)$ *Expression B:* $3x + 1$

(A) *They are equivalent — both give* 7

(B) *They are not equivalent — A gives* 9 *and B gives* 7

(C) *They are equivalent — both give* 9

(D) *They are not equivalent — A gives* 7 *and B gives* 9

16. *Emma earns $8 per hour babysitting. She also got a $15 tip. Which expression gives her total earnings for* h *hours?*

(A) $8 + 15h$

(B) $23h$

(C) $8h + 15$

(D) $8h - 15$

17. *Solve:* $8w = 72$

(A) $w = 8$

(B) $w = 64$

(C) $w = 9$

(D) $w = 80$

18. *The temperature stayed below* 0 *degrees. Which inequality represents the temperature t?*

(A) $t > 0$

(B) $t < 0$

(C) $t \leq 0$

(D) $t \geq 0$

19. *The pool opens when the temperature is more than* 75°F. *Which describes the graph of this inequality?*

(A) *Closed circle at* 75, *shade right*

(B) *Open circle at* 75, *shade right*

(C) *Closed circle at* 75, *shade left*

(D) *Open circle at* 75, *shade left*

Find more at
ViewMath.com/AL-Grade6

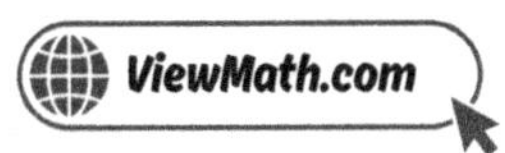

20. A triangle has base $\frac{1}{2}$ ft and height $\frac{1}{4}$ ft. What is the area?

- (A) $\frac{1}{8}$ ft^2
- (B) $\frac{3}{4}$ ft^2
- (C) $\frac{1}{16}$ ft^2
- (D) $\frac{1}{4}$ ft^2

21. A composite figure is made of a rectangle (8 cm by 5 cm) attached to a triangle with base 8 cm and height 3 cm. What is the total area?

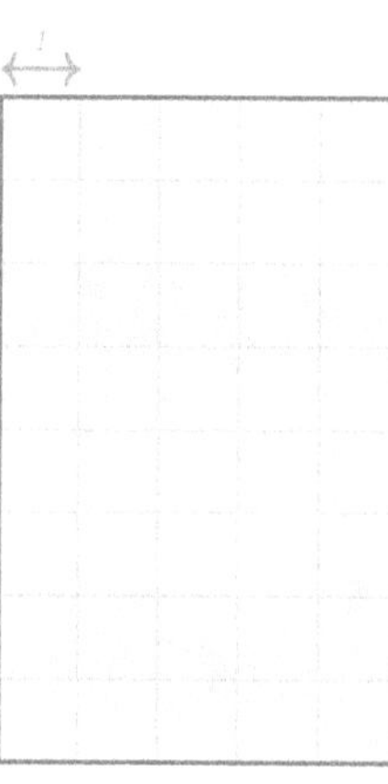

- (A) 52 cm^2
- (B) 40 cm^2
- (C) 55 cm^2
- (D) 64 cm^2

22. What is the volume of a rectangular prism with length 5 cm, width 3 cm, and height 4 cm?

- (A) 12 cm^3
- (B) 60 cm^3
- (C) 30 cm^3
- (D) 94 cm^3

23. Points $(4, 1)$ and $(-2, 1)$ form one side of a rectangle. What is the length of that side?

- (A) 2 units
- (B) 4 units
- (C) 6 units
- (D) 3 units

24. *A right triangle has vertices $(0,0)$, $(8,0)$, and $(0,6)$. What is the area?*

(A) 48 square units

(B) 24 square units

(C) 14 square units

(D) 28 square units

25. *Look at the two groups of questions below.*

Group A	Group B
How many pets do you have?	How many sides does a square have?
How tall are you in inches?	How many inches are in a foot?
How far do you live from school?	What is $3 + 4$?

Which group contains only statistical questions?

(A) Group A

(B) Group B

(C) Both groups

(D) Neither group

26. *Student test times (minutes): $8, 9, 10, 10, 11, 12, 25$. A classmate says the typical time is about 12 minutes. Is this a good estimate?*

(A) Yes, 12 is in the data set.

(B) Yes, because the outlier should be included.

(C) No, most times cluster around 9–11, so a typical value is closer to 10.

(D) No, the typical value is 25.

27. *The bar graph below shows test scores for 5 students.*

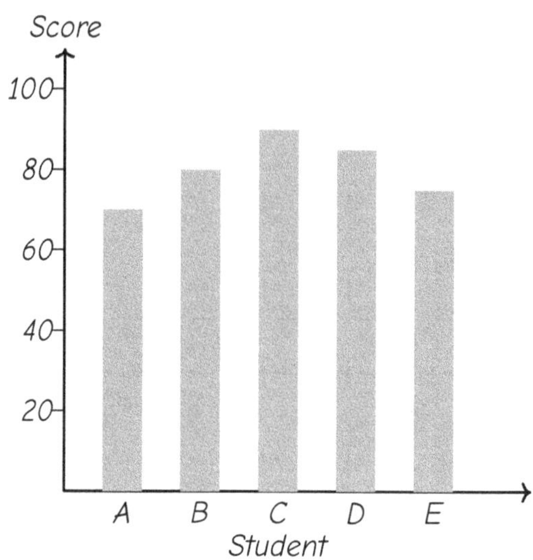

The scores are 70, 80, 90, 85, 75. What is the mean score?

(A) 75

(B) 80

(C) 85

(D) 90

28. *In a histogram, the bar for the interval 50–59 has a height of 8. What does this mean?*

(A) The values range from 50 to 59.

(B) There are 8 data values between 50 and 59.

(C) The average of the values is 8.

(D) The interval is 8 units wide.

29. *Data:* 1, 3, 5, 7, 9, 11, 13, 15, 17, 19. *What is the IQR?*

(A) 8

(B) 10

(C) 14

(D) 18

30. *Two basketball players' points per game: Player A: 18, 20, 22, 24, 26. Player B: 10, 15, 22, 28, 35. Find the mean and range for each player and compare.*

Your Answer:

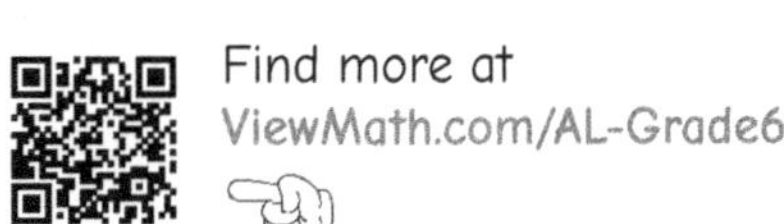
Find more at
ViewMath.com/AL-Grade6

 # End of Practice Test 9 

Great job finishing the test!

 **My Score**

I got _____________ out of 30 questions right.

Check your answers in the Answer Key at the back of the book.

Review any questions you missed. That's how we learn!

Check Your Score Online!

Visit **ViewMath Academy** to enter your answers and see which topics you need to review. You can also explore lessons, take quizzes, track your scores, and save your progress!

viewmath.com/score/6.1.AL.24

Or go to viewmath.com/score and enter code: 6.1.AL.24

10

Practice Test 10

30 Questions

✏️ Before You Start ✏️

- ✓ **Read each question carefully** before choosing your answer.
- ✓ **Show your work** on scratch paper when you need to.
- ✓ **Skip hard questions** and come back to them later.
- ✓ **Check your answers** when you're done.
- ✓ **Take your time** — there's no rush!

⭐ You've Got This! ⭐

Do your best and show what you know!

1. The tape diagram below shows the ratio of apple juice to orange juice in a drink.

Apple: ☐☐☐

Orange: ☐☐☐☐☐

If each part equals 4 ounces, how many total ounces of drink are there?

(A) 12

(B) 20

(C) 32

(D) 8

2. The table below shows the prices at two stores for packs of pencils.

	Number of Pencils	Price
Store A	10	$4.50
Store B	8	$3.20

Which store has the lower unit price per pencil?

(A) Store A at $0.45 per pencil

(B) Store B at $0.40 per pencil

(C) Store A at $0.40 per pencil

(D) They have the same unit price.

3. Which pair of ratios are equivalent?

(A) $2 : 3$ and $4 : 9$

(B) $2 : 3$ and $6 : 9$

(C) $2 : 3$ and $8 : 9$

(D) $2 : 3$ and $3 : 2$

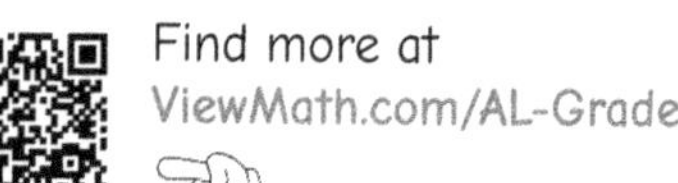
Find more at
ViewMath.com/AL-Grade6

4. *Which table matches a graph that passes through* $(0,0)$, $(2,3)$, *and* $(4,6)$?

A)
x	y
2	3
6	9

B)
x	y
2	3
6	8

C)
x	y
3	2
6	9

D)
x	y
2	4
6	12

5. *Estimate:* 48% *of* 200 *is approximately* _____.

A) 48

B) 100

C) 96

D) 400

6. *The number line below shows a distance in centimeters.*

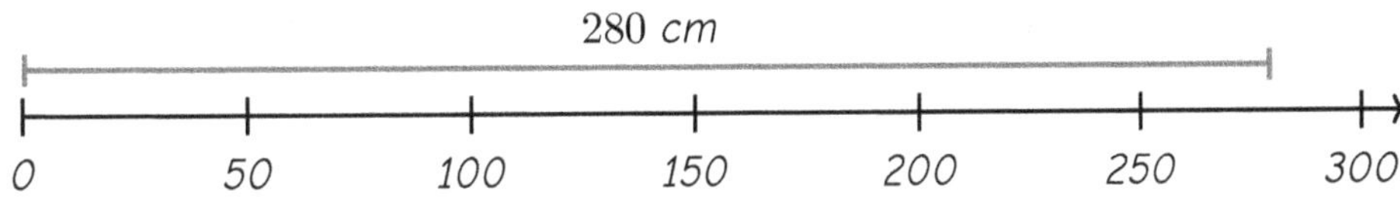

Part A: *Convert* 280 *centimeters to meters.*

Part B: *Convert* 280 *centimeters to millimeters.*

Part C: *A doorway is 2 meters tall. Is 280 cm longer or shorter? By how much?*

Your Answer:

7. *Evaluate:* $\dfrac{7}{10} \div \dfrac{2}{5}$

Your Answer:

8. Compute $6{,}048 \div 21$. Then check your answer using multiplication.

9. On a number line, where are negative numbers located?

(A) To the right of zero

(B) To the left of zero

(C) Above zero

(D) On top of zero

10. What is $|0|$?

(A) There is no answer

(B) -1

(C) 1

(D) 0

11. The area of each small square below is 1 square unit. Write the area using an exponent, then find the total area.

5 units
5 units

12. What does the expression $h + 12$ represent if h is Hannah's height in inches?

(A) Hannah's height minus 12 inches

(B) A height that is 12 inches more than Hannah's

(C) Hannah's height times 12

(D) 12 inches divided by Hannah's height

Find more at
ViewMath.com/AL-Grade6

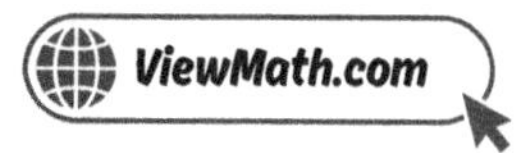

13. Look at the expression diagram below. Identify: (a) the coefficient of y, (b) the constant, and (c) the total number of terms.

$$9y \ + \ \underline{2x} \ \underline{-\ 5} \ \underline{+\ y}$$

Term 1 Term 2 Term 3 Term 4

Your Answer

14. Evaluate $10 - 2(k - 1)$ when $k = 4$.

(A) 2

(B) 4

(C) 6

(D) 14

15. Which expression is equivalent to $7(k - 3) + 5$?

(A) $7k - 16$

(B) $7k + 2$

(C) $7k - 26$

(D) $7k - 21$

16. A taxi charges \$3 plus \$2 per mile. Write an expression for the cost of a ride that is m miles long.

Your Answer

17. Solve: $x + 3.5 = 10$

(A) $x = 6.5$

(B) $x = 13.5$

(C) $x = 7.5$

(D) $x = 3.5$

18. Look at the elevator sign below. Write an inequality for the maximum number of people p and the maximum weight w.

ELEVATOR CAPACITY

Maximum: 12 persons

Maximum weight: 2,000 lbs

Your Answer:

19. Which inequality has a graph with an open circle at 0 and shading to the right?

(A) $x \leq 0$ (B) $x \geq 0$

(C) $x > 0$ (D) $x < 0$

20. A triangle has an area of $24\ ft^2$ and a base of $8\ ft$. What is the height?

(A) $3\ ft$ (B) $6\ ft$

(C) $16\ ft$ (D) $4\ ft$

21. What is the area of a parallelogram with base 9 cm and height 5 cm?

(A) $14\ cm^2$　　　　　　　　　　　　　　(B) $45\ cm^2$

(C) $22.5\ cm^2$　　　　　　　　　　　　　(D) $28\ cm^2$

22. A toy box is 3 ft long, 2 ft wide, and 2 ft tall. What is the volume?

(A) $7\ ft^3$　　　　　　　　　　　　　　　(B) $12\ ft^3$

(C) $6\ ft^3$　　　　　　　　　　　　　　　(D) $24\ ft^3$

23. Three vertices of a rectangle are $(2,3)$, $(2,-5)$, and $(9,3)$. What is the fourth vertex?

Your Answer

24. *The L-shaped figure below is drawn on a grid. What is its area?*

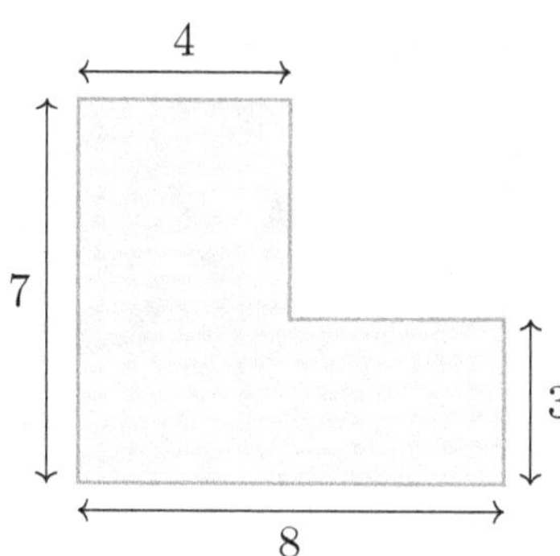

(A) 40 square units   (B) 32 square units

(C) 56 square units   (D) 24 square units

25. *A survey question is: "How many minutes does it take you to get to school?" Which is true about the data collected?*

(A) Every student will give the same answer.   (B) The answers will vary depending on each student's commute.

(C) The data cannot be displayed on a graph.   (D) The question cannot produce numerical data.

26. *Data: $60, 62, 64, 66, 68, 70$. Is this data symmetric? Explain.*

27. *What is the median of the data set $3, 7, 9, 15, 21$?*

(A) 7   (B) 9

(C) 11   (D) 15

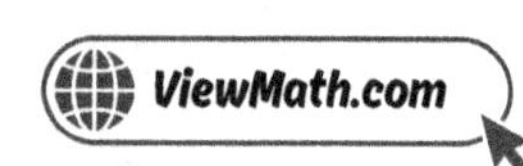

28. Look at the dot plot below showing the number of hours students studied for a test.

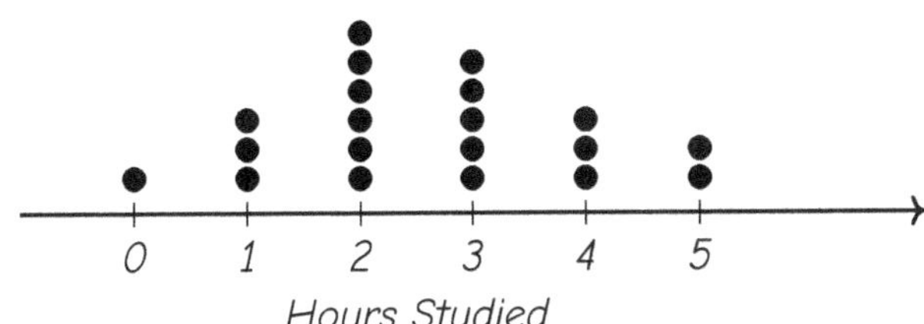

How many students studied 3 or more hours?

(A) 5

(B) 8

(C) 10

(D) 20

29. Data (in order): $8, 12, 14, 18, 22, 26, 30$. What is the five-number summary?

(A) $8, 12, 18, 26, 30$

(B) $8, 14, 18, 22, 30$

(C) $8, 12, 14, 26, 30$

(D) $8, 13, 18, 24, 30$

30. Data: $15, 18, 20, 22, 25$. The mean is 20 and the range is 10. Which is the best one-sentence summary?

(A) The data has 5 values.

(B) The typical value is about 20, and values span 10 units.

(C) The smallest value is 15.

(D) The data is not statistical.

 End of Practice Test 10 

Great job finishing the test!

My Score

I got _____________ out of 30 questions right.

Check your answers in the **Answer Key** at the back of the book

Review any questions you missed. That's how we learn!

Check Your Score Online!

Visit **ViewMath Academy** to enter your answers and see which topics you need to review. You can also explore lessons, take quizzes, track your scores, and save your progress!

viewmath.com/score/6.1.AL.25

Or go to viewmath.com/score and enter code: 6.1.AL.25

Answer Key & Explanations

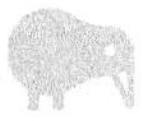

Answer Key

First try each test on your own, then check your work here.

✅ Practice Test 1 — Answer Key

1. 5
2. C
3. C
4. C
5. D
6. 8 full bottles
7. A
8. D
9. Neither positive nor negative
10. 23
11. A
12. $3n - 8$
13. $4x + y + 10$ (or any valid expression with the required features)
14. A
15. $6a + 10$
16. D
17. $n - 19 = 31;\ n = 50$
18. $n = 4$ is NOT a solution to $n > 4$ (since 4 is not greater than 4). $n = 4$ IS a solution to $n \geq 4$ (since 4 equals 4).
19. C
20. C
21. C
22. C
23. B
24. A
25. B
26. B
27. B
28. 25
29. D
30. No.

💡 Time to Learn! 💡

*Review the explanations below, **especially for the questions you missed**.*

Understanding why each answer is correct builds stronger problem-solving skills.

***Tip:** Circle any questions you got wrong, then read their explanation carefully.*

Find more at
ViewMath.com/AL-Grade6

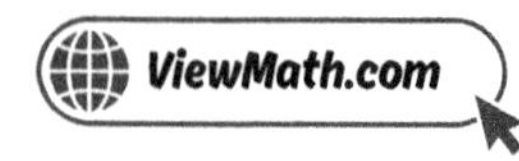

📖 Practice Test 1 — Detailed Explanations

1   *Each part = $20 \div 4 = 5$. Vegetables $= 1 \times 5 = 5$.*

2   *From the graph, the truck travels 50 miles in 2 hours. Unit rate: $50 \div 2 = 25$ miles per hour.*

3   *From the double number line, 9 scoops aligns with 15 cups of water.*

4   *$20 \div 5 = 4$. The unit rate is 4 (units of y per 1 unit of x).*

5   *Backpack: $0.15 \times \$40 = \$6$. Shoes: $0.10 \times \$60 = \$6$. Hat: $0.25 \times \$20 = \$5$. Backpack and shoes are tied at $\$6$.*

6   *$6\ L = 6{,}000\ mL$. $6{,}000 \div 750 = 8$.*

7   *The number line shows 5 jumps of $\dfrac{1}{6}$ to reach $\dfrac{5}{6}$. This models $\dfrac{5}{6} \div \dfrac{1}{6} = 5$.*

8   *$1{,}344 \div 8 = 168$ pages per week. Check: $168 \times 8 = 1{,}344$.*

9   *Zero is the boundary between positive and negative. It is neither positive nor negative.*

10   *$|-23| = 23$ because -23 is 23 units from zero on the number line. Think: how many steps does it take to get from -23 back to 0? The answer is 23 steps.*

11   *Test $n = 1$: $1^2 + 2 = 3$ ✓. $n = 2$: $4 + 2 = 6$ ✓. $n = 3$: $9 + 2 = 11$ ✓. $n = 4$: $16 + 2 = 18$ ✓.*

12   *Start with n. Multiply by 3: $3n$. Subtract 8: $3n - 8$.*

Find more at
ViewMath.com/AL-Grade6

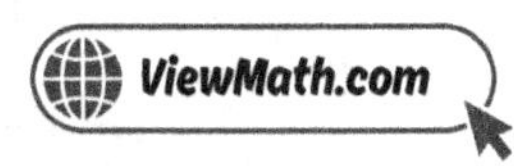

13. Answers vary. A sample: $4x + y + 10$ has 3 terms, one with coefficient 4, and constant 10.

14. $3(4) + 7 = 12 + 7 = 19$.

15. $3a + 5a - 2a = 6a$ and $7 + 3 = 10$. Result: $6a + 10$.

16. Start with $50 and subtract what you spend: $50 - d$.

17. Add 19: $n = 31 + 19 = 50$.

18. $>$ does not include equality. $\geq$ includes equality.

19. $0 > -5$ is true. Since 0 is in the shaded region (right of -5), it confirms the graph.

20. One sail: $\frac{1}{2} \times 3 \times 7 = 10.5\ m^2$. Two sails: $10.5 \times 2 = 21\ m^2$.

21. Both use $A = b \times h = 7 \times 4 = 28\ m^2$. They have equal area.

22. Volume measures 3-dimensional space, so it uses cubic units like cm^3, in^3, or ft^3.

23. Horizontal side: $|7 - 1| = 6$. Vertical side: $|2 - (-4)| = 6$. Both sides are 6 units, so it is a square. The side length is 6.

24. Rectangle: $5 \times 4 = 20$. Triangle: $\frac{1}{2} \times 5 \times 3 = 7.5$. Total: $20 + 7.5 = 27.5$ square units.

25. Different players can do different numbers of push-ups, so the data varies.

Find more at
ViewMath.com/AL-Grade6

ViewMath.com

26 Data Set A has bars that are tallest in the middle and shorter on the sides (symmetric). Data Set B has the tallest bar on the left with bars decreasing to the right (skewed right).

27 The mean and median are equal for symmetric data, but not in general. An outlier can push the mean higher or lower than the median.

28 $5 + 10 + 8 + 2 = 25$ students.

29 9 values. Median is the 5th value (40). Upper half: $45, 50, 55, 60$. $Q3 = (50 + 55) \div 2 = 52.5$.

30 Range only measures spread, not center. A smaller range means more consistency but doesn't say anything about whether scores were higher or lower. You need to compare centers too.

☑ Practice Test 2 — Answer Key

1 C

2 Part A: Lily earns $12 per hour, Noah earns $13 per hour.   Part B: Noah earns $1 more per hour.

3 Part A: 2 : 1;   Part B: 4 cups of sugar

4 C   **5** B   **6** C   **7** B   **8** 301   **9** C

10 200 feet   **11** B   **12** $\dfrac{t}{4} - 1$ **13** $A = 4, B = (x + 3)$ **14** $23   **15** A   **16** B

17 B   **18** C   **19** $x \leq -4$ **20** D   **21** A   **22** 500 m^3 **23** C   **24** 12 square units

25 B   **26** Center: about 8. Range: $10 - 5 = 5$. Shape: skewed left. Data clusters at 8–9. No major outliers.

27 B   **28** C   **29** A   **30** B

Find more at
ViewMath.com/AL-Grade6

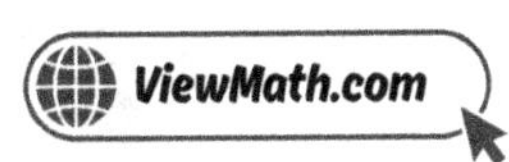

> ## 💡 Time to Learn! 💡
>
> *Review the explanations below, **especially for the questions you missed**.*
>
> *Understanding why each answer is correct builds stronger problem-solving skills.*
>
> ***Tip:** Circle any questions you got wrong, then read their explanation carefully.*

📖 Practice Test 2 — Detailed Explanations

1. *Girls have 6 parts = 24, so each part = $24 \div 6 = 4$. Boys have 4 parts = $4 \times 4 = 16$.*

2. *Lily: $\$60 \div 5 = \$12/hr$. Noah: $\$52 \div 4 = \$13/hr$. Noah earns $\$13 - \$12 = \$1$ more per hour.*

3. *Part A: From the graph, $(2, 1)$ shows 2 cups of flour for every 1 cup of sugar. Part B: Following the pattern, $(8, 4)$, so 4 cups of sugar.*

4. *At $(2, 5)$: $x : y = 2 : 5$. Check: $(4, 10)$ simplifies to $2 : 5$ as well.*

5. *$0.15 \times 4{,}000 = 600$ children.*

6. *$1\ km = 1{,}000\ m = 100{,}000\ cm$. $10 \times 100{,}000 = 1{,}000{,}000\ cm$.*

7. *$\dfrac{3}{5} \times \dfrac{5}{1} = \dfrac{15}{5} = 3$.*

8. *$45 \div 15 = 3$; bring down $1 \rightarrow 1 \div 15 = 0$ R1; bring down $5 \rightarrow 15 \div 15 = 1$. Answer: 301. The zero in the tens place is important. Check: $301 \times 15 = 4{,}515$.*

9. *Zero in a bank account means no savings and no debt. It is the boundary between having money (positive) and owing money (negative).*

Find more at
ViewMath.com/AL-Grade6

10 The distance from sea level is the absolute value of the position: $|-200| = 200$. The submarine is 200 feet from sea level.

11 Parentheses: $2 + 4 = 6$. Exponent: $6^2 = 36$. Multiply: $3 \times 36 = 108$.

12 Divide t by 4 to get $\frac{t}{4}$, then subtract 1: $\frac{t}{4} - 1$.

13 $4(x + 3) = 4 \times (x + 3)$. The two factors are 4 and $(x + 3)$.

14 $5 + 3(6) = 5 + 18 = 23$ dollars.

15 Combine like terms: $6a + 2a = 8a$ and $3 - 1 = 2$. Result: $8a + 2$.

16 A square has 4 equal sides, so its perimeter is 4 times the side length: $P = 4s$.

17 Subtract 15: $k = 32 - 15 = 17$.

18 $n \leq 8$ means n is 8 or less. $8 \leq 8$ is true. 8.5, 9, and 10 are all greater than 8.

19 Closed circle means $\leq$ or $\geq$. Shading left means less than or equal to: $x \leq -4$.

20 $A = \frac{1}{2} \times 14 \times 8 = 56$ m^2.

21 $A = 8 \times 3.5 = 28$ m^2.

22 $V = 25 \times 10 \times 2 = 500$ m^3.

23 Vertical segment: $|7 - (-1)| = |8| = 8$ units.

Find more at
ViewMath.com/AL-Grade6

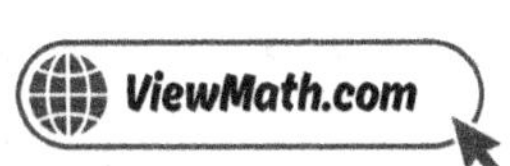

24 Rectangle area $= 6 \times 4 = 24$. Triangle area $= \frac{1}{2} \times 6 \times 4 = 12$. Remaining $= 24 - 12 = 12$ square units.

25 The dot plot shows data ranging from 0 to 5, with different students giving different answers. This variability confirms the question is statistical.

26 The peak is at score 8 with 5 dots. Most scores are at 7–10. The tail stretches to the left (5 and 6), so the data is skewed left. Range $= 5$.

27 Mean $= (4 + 6 + 8 + 10 + 12) \div 5 = 40 \div 5 = 8$.

28 There are no data values between 25 and 30, creating a gap in the data.

29 Class A's box (IQR) is narrower than Class B's. A narrower box means the middle 50% of scores are closer together — more consistent.

30 Class X ranges from 25 to 40 (range $= 15$). Class Y ranges from 15 to 50 (range $= 35$). Class Y is much more spread out.

📋 Practice Test 3 — Answer Key

1 C   **2** B   **3** B   **4** B   **5** C   **6** C   **7** A   **8** C   **9** C   **10** D

11 D   **12** $\dfrac{a+b}{2}$ **13** A   **14** 5   **15** A   **16** C   **17** C   **18** D   **19** D   **20** C

21 B   **22** 84 cubic units   **23** A   **24** B   **25** C

26 Skewed to the left (or: most data on the right with a tail to the left).   **27** C   **28** B   **29** B

30 B

Find more at
ViewMath.com/AL-Grade6

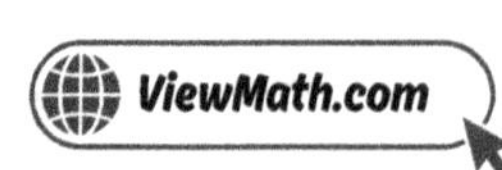

> 💡 *Time to Learn!* 💡
>
> *Review the explanations below, **especially for the questions you missed**.*
>
> *Understanding why each answer is correct builds stronger problem-solving skills.*
>
> ***Tip:*** *Circle any questions you got wrong, then read their explanation carefully.*

📖 Practice Test 3 — Detailed Explanations

1 The question asks flour to eggs. Flour $= 3$, eggs $= 2$. The ratio is $3 : 2$.

2 $\$9 \div 12 = \$0.75$ per muffin.

3 $7 \times 3 = 21$ bananas, so $\$2 \times 3 = \$6$.

4 The ratio is $6 : 2 = 3 : 1$. When $x = 15$: $y = 15 \div 3 = 5$.

5 $0.05 \times 240 = 12$.

6 $3 \times 5{,}280 = 15{,}840$ feet.

7 $\dfrac{1}{3} \times \dfrac{3}{2} = \dfrac{3}{6} = \dfrac{1}{2}$.

8 $15 \div 5 = 3$; bring down $7 \rightarrow 7 \div 5 = 1$ R2; bring down $5 \rightarrow 25 \div 5 = 5$. Answer: 315. Check: $315 \times 5 = 1{,}575$.

9 Above sea level is represented by a positive integer. $1{,}200$ feet above sea level is $+1{,}200$ or simply $1{,}200$.

10 The opposite of 5 is -5, which is written as $-(5)$. Choice A gives $|5| = 5$. Choice C gives $-(-5) = 5$ (the opposite of the opposite is the original number).

11 The base 5 is multiplied 3 times: 5^3.

12 First find the sum $a + b$, then divide by 2: $\frac{a+b}{2}$.

13 $7k$ is a single term. There are no $+$ or $-$ signs separating other pieces.

14 $(30 - 5) \div 5 = 25 \div 5 = 5$.

15 GCF of 12 and 18 is 6. Factor: $6(2x + 3)$. Check: $6 \times 2x + 6 \times 3 = 12x + 18$. ✓

16 $50 + 40(3) = 50 + 120 = 170$ dollars.

17 Step 1 is correct (multiply to undo division), but Step 2 should be $n = 8 \times 5 = 40$, not $8 + 5 = 13$.

18 "No more than 200" means 200 or fewer. $c \leq 200$.

19 $\leq$ includes -1 (closed circle). Less than -1 is to the left (shade left).

20 $A = \frac{1}{2} \times 5 \times 12 = 30$ in^2.

21 $A = \frac{1}{2}(9 + 15)(6) = \frac{1}{2}(24)(6) = 72$ in^2.

22 Bottom prism: $8 \times 3 \times 2 = 48$. Top prism: $3 \times 3 \times 4 = 36$. Total: $48 + 36 = 84$ cubic units.

23 From $(-2, -2)$, moving 5 right gives $x = 3$, and 5 up gives $y = 3$. The opposite vertex is $(3, 3)$.

Find more at
ViewMath.com/AL-Grade6

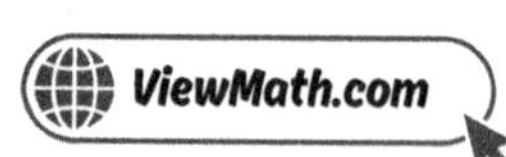

24 Length $= |3 - (-5)| = 8$. Width $= |4 - (-1)| = 5$. Area $= 8 \times 5 = 40$. Jake is correct.

25 On a single day, the total is one number. However, if she collects data across many days, the number would vary and become statistical. As asked (for one day), it has a single answer.

26 The peak is at 5, with values trailing to the left toward 1. Most data is on the higher end.

27 Total $= 15 \times 6 = 90$. Remove 9: $90 - 9 = 81$. New mean $= 81 \div 5 = 16.2$.

28 Total $= 2 + 5 + 4 + 3 + 1 = 15$ students.

29 The box represents $Q1$ to $Q3$, which contains the middle 50% of the data.

30 Little overlap means most values in one group are different from the values in the other group, which signals a clear difference between the groups.

✅ Practice Test 4 — Answer Key

1 B   2 3.5 hours   3 C   4 Line A. It has a rate of 6 per 1, which is higher than 4 per 1.

5 B   6 D   7 A   8 168 cases   9 C   10 C   11 A   12 $3n + 2$ 13 B

14 25   15 A   16 A   17 C   18 C   19 B   20 C   21 B   22 A   23 B

24 64 square units   25 C   26 B   27 B   28 C

29 Store B has higher typical sales (median $\approx$ \$275 vs. \$250). Store A has more predictable sales (IQR $\approx$ \$100 vs. \$125).

30 B

Find more at
ViewMath.com/AL-Grade6

💡 Time to Learn! 💡

*Review the explanations below, **especially for the questions you missed**.*

Understanding why each answer is correct builds stronger problem-solving skills.

Tip: *Circle any questions you got wrong, then read their explanation carefully.*

📖 Practice Test 4 — Detailed Explanations

1. "3 pencils for every 1 eraser" means pencils to erasers is $3 : 1$.

2. $49 \div 14 = 3.5$ hours.

3. The ratio is $4 : 10 = 2 : 5$. Row 3 should be $12 : 30$ (since $4 \times 3 = 12$ and $10 \times 3 = 30$), but it shows $12 : 25$.

4. Line A goes up 6 for every 1 unit right; Line B goes up 4. Higher rise per unit means a faster rate.

5. $0.30 \times 90 = 27$.

6. $3 \times 4 = 12$ cups.

7. Dividing by $\frac{c}{d}$ is the same as multiplying by its reciprocal, $\frac{d}{c}$.

8. $5{,}376 \div 32 = 168$. Check: $168 \times 32 = 5{,}376$.

9. A negative number is less than zero. -4 is the only number less than zero in the list. Zero is neither positive nor negative.

Find more at
ViewMath.com/AL-Grade6

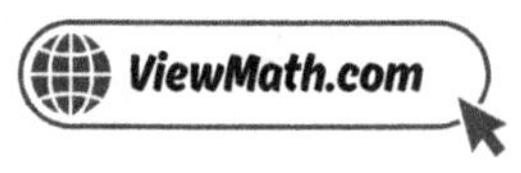

10 Absolute value is the distance from zero, which is always positive or zero. $|-8| = 8$ because -8 is 8 units from zero on the number line.

11 Multiply: $4 \times 3 = 12$. Then left to right: $20 - 12 + 1 = 9$.

12 n notebooks cost $3n$ dollars. Add the pen: $3n + 2$.

13 In $r - 5$, the coefficient of r is 1 (since $r = 1 \cdot r$), not 0. Row 2 is incorrect.

14 $4^2 + 2(4) + 1 = 16 + 8 + 1 = 25$.

15 $4(n + 2) = 4n + 8$ by the distributive property. They are equivalent for every value of n.

16 New fencing: top (12) + right (w) + bottom $(12) = 12 + w + 12 = 24 + w$.

17 Divide by 12: $p = 84 \div 12 = 7$.

18 "More than \$25" means greater than 25: $d > 25$.

19 $x < -3$: $-4 < -3$ is true (shaded), but $-2 < -3$ is false (not shaded).

20 $A = \frac{1}{2} \times 20 \times 7 = 70$ in^2.

21 Area uses the height, not the slant side. $A = 12 \times 6 = 72$ m^2.

22 $V = \frac{1}{2} \times \frac{1}{2} \times \frac{1}{2} = \frac{1}{8}$ m^3.

23 $A(1, -2)$ and $B(7, -2)$ have the same y-coordinate. Distance $= |7 - 1| = 6$ units.

Find more at
ViewMath.com/AL-Grade6

24   Bottom rectangle: $10 \times 4 = 40$. Top-left rectangle: $6 \times (8 - 4) = 6 \times 4 = 24$. Total: $40 + 24 = 64$ square units.

25   A statistical question is one where you expect different answers from different people or situations.

26   A larger spread means values are more scattered from the center, even if the center values are identical.

27   Original mean: $(58 + 60 + 62 + 64) \div 4 = 61$. New mean: $(58 + 60 + 62 + 64 + 82) \div 5 = 326 \div 5 = 65.2$.

28   Total $= 3 + 7 + 10 + 5 = 25$.

29   Store A: median $\approx \$250$, IQR $\approx 300 - 200 = \$100$. Store B: median $\approx \$275$, IQR $\approx 350 - 225 = \$125$. Store B sells more on a typical day, but Store A's sales are more consistent (smaller IQR and range).

30   Runner B has a lower mean time ($24 < 25$), so faster. Runner A has a smaller MAD ($1 < 5$), so more consistent.

✅ Practice Test 5 — Answer Key

1   10   **2**   C   **3**   C   **4**   B   **5**   B   **6**   150 *minutes;* 9,000 *seconds*   **7**   C   **8**   B

9   C   **10**   11 *and* -11 **11**   39   **12**   C   **13**   5 *and* $(y + 8)$ **14**   B   **15**   C

16 $100 - 8h$ **17**   D   **18**   C   **19**   C   **20**   B   **21**   7 cm   **22**   A   **23**   A   **24**   C

25   C   **26**   A   **27**   Mean $= 23.75$, Median $= 23.5$ **28**   B

29   Min $= 2$, Q1 $= 4.5$, Median $= 8$, Q3 $= 11$, Max $= 15$. IQR $= 6.5$.   **30**   B

Find more at
ViewMath.com/AL-Grade6

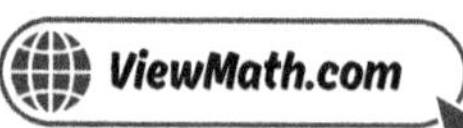

💡 Time to Learn! 💡

*Review the explanations below, **especially for the questions you missed**.*

Understanding why each answer is correct builds stronger problem-solving skills.

***Tip:** Circle any questions you got wrong, then read their explanation carefully.*

📖 Practice Test 5 — Detailed Explanations

1. *Total parts* $= 3 + 2 = 5$. *Each part* $= 25 \div 5 = 5$. *Swimming* $= 2 \times 5 = 10$.

2. $240 \div 8 = 30$ *miles per gallon.*

3. $4 \times 2.5 = 10$ *loaves, so* $6 \times 2.5 = 15$ *cups. Alternatively, unit rate:* $6 \div 4 = 1.5$ *cups per loaf, then* $1.5 \times 10 = 15$.

4. *Ratio:* $1 : 4$. *When* $x = 3$: $y = 3 \times 4 = 12$.

5. $0.50 \times ? = 33$. *Divide:* $33 \div 0.50 = 66$.

6. $2 \times 60 + 30 = 150$ *minutes.* $150 \times 60 = 9{,}000$ *seconds.*

7. $\dfrac{2}{3} \times \dfrac{5}{4} = \dfrac{10}{12} = \dfrac{5}{6}$.

8. *The four steps of the standard algorithm are **Divide, Multiply, Subtract, Bring Down**. After dividing, you multiply the partial quotient by the divisor before subtracting.*

9. *A gain of yards is a positive change. Debt, below sea level, and below zero are all represented by negative numbers.*

Find more at
ViewMath.com/AL-Grade6

(10) Both 11 and -11 have an absolute value of 11 because each is 11 units from zero. $|11| = 11$ and $|-11| = 11$.

(11) $7^2 = 49$. Then $49 - 10 = 39$.

(12) "The quotient of m and 5" means m divided by 5: $m \div 5$.

(13) $5(y + 8) = 5 \times (y + 8)$. The factors are 5 and $(y + 8)$.

(14) $2^3 = 2 \times 2 \times 2 = 8$.

(15) Distribute: $5 \times 2m + 5 \times 3 = 10m + 15$.

(16) Start at 100 and subtract 8 per hour: $100 - 8h$.

(17) Division is undone by multiplication. Multiply both sides by 3: $n = 12 \times 3 = 36$.

(18) $n \geq 12$ means n is 12 or more, which is "at least 12."

(19) $-3 \geq -3$ is true (they are equal). The other values are less than -3.

(20) $P = \frac{1}{2} \times 10 \times 6 = 30$. $Q = \frac{1}{2} \times 10 \times 8 = 40$. Difference $= 40 - 30 = 10$ cm^2.

(21) $84 = \frac{1}{2}(10 + 14)h = \frac{1}{2}(24)h = 12h$. So $h = 84 \div 12 = 7$ cm.

(22) Base area $= 10 \times 4 = 40$. Height $= 120 \div 40 = 3$ cm.

(23) From $(0,0)$ with length 10 along the x-axis and width 4 along the y-axis, the opposite vertex is $(10, 4)$.

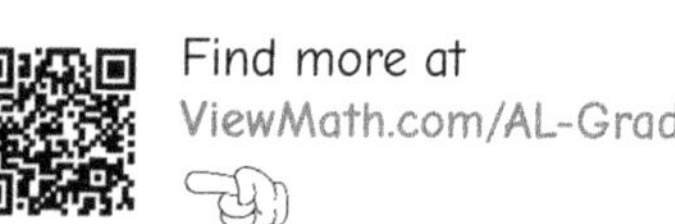
Find more at
ViewMath.com/AL-Grade6

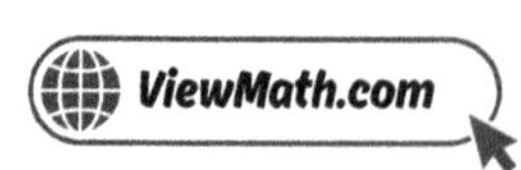

24. Length $= 6$, width $= 4$. Area $= 6 \times 4 = 24$ square units.

25. Different students read different numbers of pages, so the data varies.

26. The outlier 100 increases the mean significantly. Without it, the center is around 17. With it, the mean jumps to about 27.

27. Mean: $(18 + 22 + 25 + 30) \div 4 = 95 \div 4 = 23.75$. Median: $(22 + 25) \div 2 = 23.5$.

28. Most data clusters between 40 and 50. The values 10 and 80 are far from the cluster and may be outliers.

29. 9 values, already ordered. Median is the 5th value: 8. Lower half: $2, 4, 5, 7$. $Q1 = (4 + 5) \div 2 = 4.5$. Upper half: $9, 10, 12, 15$. $Q3 = (10 + 12) \div 2 = 11$. $IQR = 11 - 4.5 = 6.5$.

30. The range is 12 and the data is evenly spaced. Whether "very spread out" is accurate depends on context, but the data is quite orderly and not wildly spread.

☑ Practice Test 6 — Answer Key

1. Bananas to yogurt: $3 : 4$; Yogurt to bananas: $4 : 3$
2. B
3. B
4. A
5. B
6. B
7. C
8. B
9. B
10. B
11. B
12. A
13. 1
14. C
15. A
16. D
17. $k = 9$
18. 6, 7, and 10 (or any three numbers that are 6 or greater)
19. B
20. $\frac{1}{4}$ ft^2
21. B
22. 4 cm
23. B
24. B
25. B
26. C
27. B
28. C
29. B
30. C

Find more at
ViewMath.com/AL-Grade6

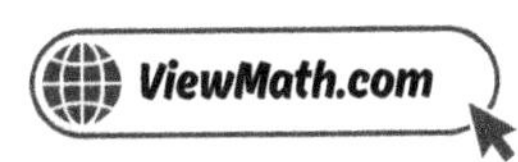

💡 Time to Learn! 💡

*Review the explanations below, **especially for the questions you missed**.*

Understanding why each answer is correct builds stronger problem-solving skills.

Tip: *Circle any questions you got wrong, then read their explanation carefully.*

📖 Practice Test 6 — Detailed Explanations

1. *"3 bananas for every 4 cups of yogurt" gives $3 : 4$. Flip the order for yogurt to bananas: $4 : 3$.*

2. *Brand X: $\$8.75 \div 5 = \$1.75$ each. Brand Y: $\$4.50 \div 3 = \$1.50$ each. Brand Y is cheaper.*

3. *$2 \times 8 = 16$ students, so $3 \times 8 = 24$ pencils.*

4. *Hours go on the x-axis, miles on the y-axis: $(1, 30)$, $(2, 60)$, $(3, 90)$.*

5. *Increase: $0.10 \times \$25 = \$2.50$. New price: $\$25 + \$2.50 = \$27.50$.*

6. *1 kg = 1,000 g. $7,000 \div 1,000 = 7$ kg.*

7. *$4 \div \dfrac{2}{3} = \dfrac{4}{1} \times \dfrac{3}{2} = \dfrac{12}{2} = 6$ batches.*

8. *After $36 \div 12 = 3$, bringing down 1 gives $1 \div 12 = 0$. The student skipped that zero and jumped to $12 \div 12 = 1$, writing 31 instead of 301.*

9. *A negative temperature means below zero. $-12°F$ means 12 degrees below zero.*

Find more at
ViewMath.com/AL-Grade6

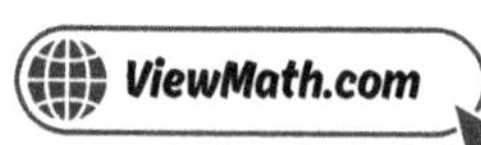

10 Absolute value tells you how far a number is from zero. -12 is 12 units from zero, so $|-12| = 12$. A common mistake is to think $|-12| = -12$, but distance is never negative.

11 $2^3 = 8$ and $3^2 = 9$. Sum: $8 + 9 = 17$.

12 There are 3 boxes labeled x and 2 boxes labeled 1. The expression is $3x + 2$.

13 $4(a + b)$ is a single term — it is one product. (If you distribute it to $4a + 4b$, it would be 2 terms.)

14 $4(3 + 2) = 4(5) = 20$.

15 $9w - 4w = 5w$ and $6 - 6 = 0$. Result: $5w$.

16 Splitting \$14 equally among f friends means dividing: $14 \div f$.

17 $k = 54 \div 6 = 9$. Check: $6(9) = 54$. ✓

18 Any number greater than or equal to 6 is a solution.

19 Both shade to the right. The only difference is the circle at 5: open for $>$, closed for $\geq$.

20 $A = \frac{1}{2} \times \frac{3}{4} \times \frac{2}{3} = \frac{1}{2} \times \frac{6}{12} = \frac{1}{2} \times \frac{1}{2} = \frac{1}{4}$ ft^2.

21 When both bases are equal ($b_1 = b_2 = 4$), the trapezoid is actually a parallelogram. Area $= \frac{1}{2}(4 + 4)(6) = 24 = 4 \times 6$.

22 Base area $= 9 \times 5 = 45$. Height $= 180 \div 45 = 4$ cm.

23 $(5, -1)$ and $(5, 4)$ share the same x-coordinate, so the segment is vertical.

24 Decompose irregular shapes into simpler shapes (rectangles and triangles) whose areas you can compute, then add them.

25 There are always 3 feet in one yard. This is not a statistical question because the answer does not vary.

26 There is a gap between 22 and 35 (no values from 23 to 34). The other data sets have values close together.

27 Sum $= 88 + 92 + 76 + 84 = 340.$ Mean $= 340 \div 4 = 85.$

28 Range $= 78 - 70 = 8.$

29 $IQR = Q3 - Q1 = 40 - 20 = 20.$

30 A complete summary uses center (mean or median) to describe what is typical and spread (range, IQR, or MAD) to describe consistency.

✅ Practice Test 7 — Answer Key

1 6   2 B   3 16 and 20   4 C   5 $64.80   6 C   7 4   8 A   9 B

10 C   11 41   12 D   13 C   14 C   15 $12x - 5$ 16 C   17 B   18 C

19 D   20 13.5 m^2 21 D   22 D   23 12 units   24 B

25 Answers vary. Example: "How many players are on a basketball team on the court at one time?"   26 B

27 Mean $= 19$, Median $= 17.$ The median is better.   28 C   29 B   30 B

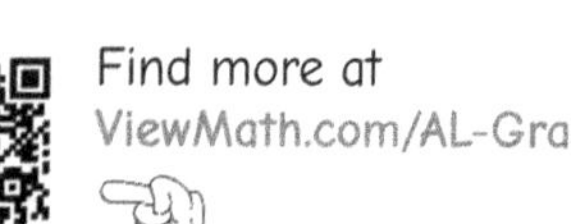
Find more at
ViewMath.com/AL-Grade6

💡 *Time to Learn!* 💡

*Review the explanations below, **especially for the questions you missed**.*

Understanding why each answer is correct builds stronger problem-solving skills.

***Tip:** Circle any questions you got wrong, then read their explanation carefully.*

📖 Practice Test 7 — Detailed Explanations

1. $21 \div 7 = 3$, so multiply both by 3: iced tea $= 2 \times 3 = 6$.

2. $150 \div 5 = 30$ gallons per hour.

3. Row 2: $5 \times 2 = 10$, so $8 \times 2 = 16$. Row 3: $8 \times 4 = 32$, so $5 \times 4 = 20$.

4. $5 : 15$ simplifies to $1 : 3$.

5. Tax $= 0.08 \times \$60 = \$4.80$. Total $= \$60 + \$4.80 = \$64.80$.

6. 1 foot $= 12$ inches. $5 \times 12 = 60$ inches.

7. $2\frac{1}{2} = \frac{5}{2}$. Then $\frac{5}{2} \div \frac{5}{8} = \frac{5}{2} \times \frac{8}{5} = \frac{40}{10} = 4$ shelves.

8. $47 \div 12 = 3$ R11; bring down 5 $\rightarrow$ $115 \div 12 = 9$ R7; bring down 2 $\rightarrow$ $72 \div 12 = 6$. Answer: 396. Check: $396 \times 12 = 4{,}752$.

9. Below sea level is represented by a negative number. 400 feet below sea level is -400.

10 $|6| = 6$ and $|-6| = 6$. Both 6 and -6 are 6 units from zero, so both have an absolute value of 6.

11 $5^2 = 25$ and $4^2 = 16$. Sum: $25 + 16 = 41$.

12 Giving away 6 means subtracting: $x - 6$.

13 The constant is the term without a variable: 10. The coefficient of x is 2, of y is 5, and there are 3 terms.

14 $3 + 2(7) = 3 + 14 = 17$ dollars.

15 $9x + 3x = 12x$ and $2 - 7 = -5$. Result: $12x - 5$.

16 Area $=$ length $\times$ width $= l \times 5 = 5l$.

17 Subtract 9 from both sides: $x = 15 - 9 = 6$.

18 $t \leq 30$ means 30 or less, which is "no more than 30 degrees."

19 The graph represents $x \leq -1$. The value $0 > -1$, so 0 is not a solution.

20 $A = \frac{1}{2} \times 4.5 \times 6 = 13.5 \ m^2$.

21 $A = \frac{1}{2}(40 + 60)(30) = \frac{1}{2}(100)(30) = 1{,}500 \ m^2$.

22 $V = 20 \times 10 \times 12 = 2{,}400 \ in^3$.

23 $|5 - (-7)| = |12| = 12$ units.

Find more at
ViewMath.com/AL-Grade6

ViewMath.com

24. Base $= 9 - 1 = 8$. Height $= 7 - 1 = 6$. Area $= \frac{1}{2} \times 8 \times 6 = 24$ square units.

25. A non-statistical question has one fixed answer. The number of players on the court (5) does not vary.

26. Data set A range: $24 - 20 = 4$. Data set B range: $34 - 10 = 24$. Data set B is much more spread out.

27. Ordered: $12, 14, 15, 16, 18, 20, 22, 35$. Sum $= 152$. Mean $= 152 \div 8 = 19$. Median $= (16 + 18) \div 2 = 17$. The 35-point game is an outlier that raises the mean. The median (17) better reflects typical performance.

28. A dot plot places one dot for each data value, so you can see every individual value.

29. Short whiskers and a narrow box indicate small range and small IQR, meaning data values are close together.

30. Group Q has a higher median ($60 > 45$) and a smaller IQR ($5 < 20$), so it is both higher and more consistent.

☑ Practice Test 8 — Answer Key

1. B

2. Part A: 0.5 laps per minute (or 1 lap every 2 minutes).   Part B: 7 laps.

3. 15

4. B

5. 360

6. C

7. B

8. C

9. B

10. $|-3|$ is equal to $|3|$

11. C

12. B

13. 12 and k

14. 30 square cm

15. $11m + 2$

16. $20r + 15$

17. $w = 72$

18. C

19. No. Both shade to the left, but $x < 8$ has an open circle at 8 and $x \leq 8$ has a closed circle at 8.

20. A

21. $27\ m^2$

22. C

23. C

24. 42 square units

25. The teacher has one specific age, so the answer does not vary.

26. B

27. Mean ≈ 13.3, Median $= 6$. The median is better.

28. B

29. Range $= 55$, IQR $= 30$

30. A

💡 Time to Learn! 💡

*Review the explanations below, **especially for the questions you missed**.*

Understanding why each answer is correct builds stronger problem-solving skills.

Tip: *Circle any questions you got wrong, then read their explanation carefully.*

📖 Practice Test 8 — Detailed Explanations

1. Dogs have 4 parts. Each part = 2 animals. Dogs = $4 \times 2 = 8$.

2. Part A: From the graph, the swimmer completes 1 lap every 2 minutes, so the rate is $\frac{1}{2}$ lap per minute. Part B: $14 \div 2 = 7$ laps.

3. $2 \times 5 = 10$ strawberries, so $3 \times 5 = 15$ cups of yogurt.

4. x doubled from 4 to 8, so y doubles from 7 to 14.

5. $0.72 \times 500 = 360$.

6. 1 yard = 3 feet. $7 \times 3 = 21$ feet.

7. $\frac{2}{7} \times \frac{6}{5} = \frac{12}{35}$. Kai is correct.

8. $73 \div 24 = 3$ R1; bring down $4 \to 14 \div 24 = 0$ R14; bring down $4 \to 144 \div 24 = 6$. Answer: 306. The zero in the tens place must not be skipped. Check: $306 \times 24 = 7{,}344$.

9. Below sea level is negative. Sea level is 0, so 75 feet below is -75.

Find more at
ViewMath.com/AL-Grade6

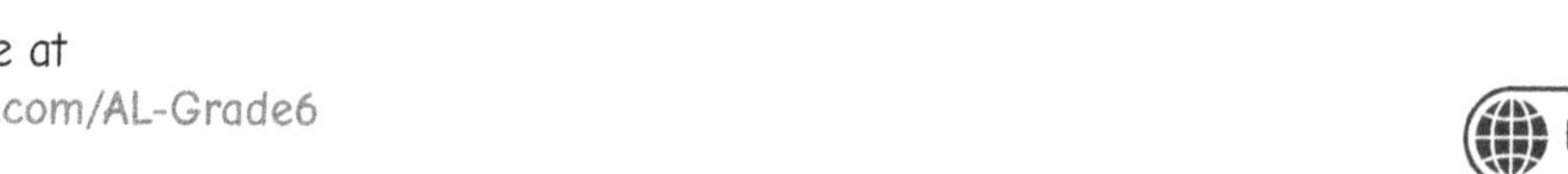

10. $|-3| = 3$ and $|3| = 3$, so they are equal. Both -3 and 3 are the same distance (3 units) from zero. A number and its opposite always have the same absolute value.

11. $3^4 = 3 \times 3 \times 3 \times 3 = 81$. Each value is 3 times the previous one: $27 \times 3 = 81$.

12. The product of 4 and t is $4t$. The sum of 9 and that product is $9 + 4t$.

13. $12k = 12 \times k$. The factors are 12 and k.

14. $A = \frac{1}{2} \times 10 \times 6 = \frac{1}{2} \times 60 = 30$ square cm.

15. Distribute: $8m + 2 + 3m$. Combine: $8m + 3m = 11m$. Result: $11m + 2$.

16. Regular seats: $20r$. VIP seats: 15. Total: $20r + 15$.

17. Multiply by 8: $w = 9 \times 8 = 72$.

18. You can drive at 65 or below, so ≤ 65.

19. The direction is the same, but the circle type differs: open vs. closed.

20. Triangle A: $\frac{1}{2} \times 4 \times 4 = 8$. Triangle B: $\frac{1}{2} \times 5 \times 3 = 7.5$. Triangle A has the greater area.

21. $A = 4.5 \times 6 = 27 \ m^2$.

22. Volume of a rectangular prism is $V = l \times w \times h$.

23. When two points have the same y-coordinate, the segment between them is horizontal.

Find more at
ViewMath.com/AL-Grade6

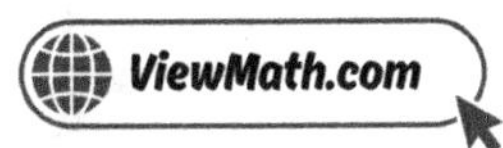

24. Length $= |9 - 2| = 7$. Width $= |5 - (-1)| = 6$. Area $= 7 \times 6 = 42$ square units.

25. A statistical question expects data that varies. The teacher's age is a single fixed number.

26. Most scores cluster around 70–76. The center is about 73–75. The 98 is an outlier that would pull the mean higher, but the typical values are in the low-to-mid 70s.

27. Mean: $(3 + 4 + 5 + 6 + 7 + 8 + 60) \div 7 = 93 \div 7 \approx 13.3$. Median: 6. The outlier 60 inflates the mean. The median (6) is more typical.

28. With 200 data points, a dot plot would be too crowded. A histogram groups the data into intervals for a clearer display.

29. Range $= 80 - 25 = 55$. IQR $= 65 - 35 = 30$.

30. Class A's MAD is 3, meaning scores are typically within 3 of the mean. Class B's scores typically deviate 9 points. Class A's student is more likely to be near 75.

☑ Practice Test 9 — Answer Key

1. C 2. B 3. B 4. B 5. B 6. C 7. D 8. 1 9. D

10. Part A: $|A| = 7$, $|B| = 3$, $|C| = 4$; Part B: Point A; Part C: Point B 11. B 12. C 13. C

14. B 15. B 16. C 17. C 18. B 19. B 20. C 21. A 22. B 23. C

24. B 25. A 26. C 27. B 28. B 29. B

30. Both have mean $= 22$. Player A range $= 8$; Player B range $= 25$. Same average, but Player A is more consistent.

Find more at
ViewMath.com/AL-Grade6

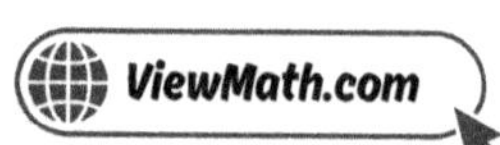

💡 *Time to Learn!* 💡

*Review the explanations below, **especially for the questions you missed.***

Understanding why each answer is correct builds stronger problem-solving skills.

Tip: *Circle any questions you got wrong, then read their explanation carefully.*

📖 Practice Test 9 — Detailed Explanations

1 *Total parts* $= 1 + 3 + 2 = 6$. *Each part* $= 18 \div 6 = 3$. *Blue* $= 3 \times 3 = 9$.

2 *Divide:* $120 \div 4 = 30$ *pages per minute.*

3 $3 \times 2 = 6$, *so* $5 \times 2 = 10$.

4 $1 : 3$ *multiplied by 3 gives* $3 : 9$, *so* $(3, 9)$ *is on the line.*

5 *Discount:* $0.20 \times \$150 = \$30$. *Sale price:* $\$150 - \$30 = \$120$.

6 $1\ L = 1{,}000\ ml$. $2.5 \times 1{,}000 = 2{,}500\ mL$.

7 $\dfrac{4}{5} \div \dfrac{2}{5} = \dfrac{4}{5} \times \dfrac{5}{2} = \dfrac{20}{10} = 2$ *servings.*

8 *After subtracting 15, the remainder is 0. Bring down 7 to get 7. Since* $7 \div 5 = 1$ *R2, the missing digit in the quotient is 1. The full answer is 315.*

9 *A withdrawal removes money, creating a negative change.* -15 *represents losing or going below by 15. A deposit, a temperature above zero, and an elevation above sea level are all positive.*

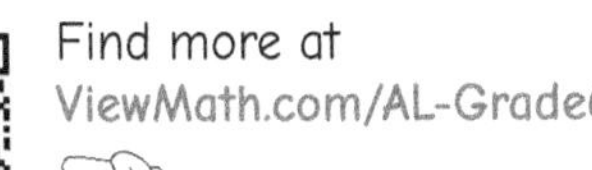
Find more at
ViewMath.com/AL-Grade6

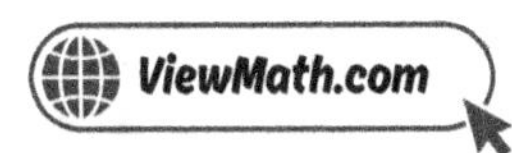

10. Part A: $|-7| = 7$, $|3| = 3$, $|-4| = 4$. Part B: Point A has the greatest absolute value (7), so it is farthest from zero. Part C: Point B has the smallest absolute value (3), so it is closest to zero.

11. Parentheses: $5 - 1 = 4$. Exponent: $4^2 = 16$. Add: $16 + 7 = 23$.

12. "Plus" means addition: $n + 8$.

13. The number in front of w is 4, making 4 the coefficient. 12 and 3 are constants. There are 3 terms.

14. $6^2 - 5 = 36 - 5 = 31$.

15. A: $3(2 + 1) = 3(3) = 9$. B: $3(2) + 1 = 7$. Different values, so not equivalent.

16. Hourly earnings: $8h$. Plus tip: $8h + 15$.

17. Divide both sides by 8: $w = 72 \div 8 = 9$.

18. "Below 0" means less than 0: $t < 0$. (Not equal to 0, since it says "below.")

19. "More than 75" is $t > 75$: open circle (not including 75), shade right (greater values).

20. $A = \frac{1}{2} \times \frac{1}{2} \times \frac{1}{4} = \frac{1}{16}$ ft^2.

21. Rectangle: $8 \times 5 = 40$. Triangle: $\frac{1}{2} \times 8 \times 3 = 12$. Total: $40 + 12 = 52$ cm^2.

22. $V = 5 \times 3 \times 4 = 60$ cm^3.

23. Same y-coordinate. $|4 - (-2)| = 6$ units.

24   Base $= 8$, height $= 6$. Area $= \frac{1}{2} \times 8 \times 6 = 24$ square units.

25   Group A questions all produce answers that vary from person to person. Group B questions each have a single fixed answer.

26   Most data values are between 8 and 12, clustering around 10. The 25 is an outlier. A typical time is about 10 minutes.

27   Mean $= (70 + 80 + 90 + 85 + 75) \div 5 = 400 \div 5 = 80$.

28   The height of a histogram bar tells you how many data values fall in that interval. A height of 8 means 8 values.

29   10 values. Median $= (9 + 11) \div 2 = 10$. Lower half: $1, 3, 5, 7, 9$. $Q1 = 5$. Upper half: $11, 13, 15, 17, 19$. $Q3 = 15$. $IQR = 15 - 5 = 10$.

30   Player A: $(18 + 20 + 22 + 24 + 26) \div 5 = 22$, range $= 8$. Player B: $(10 + 15 + 22 + 28 + 35) \div 5 = 22$, range $= 25$. Both average 22 points, but Player A's scores are much more consistent.

☑ Practice Test 10 — Answer Key

1   C   2   B   3   B   4   A   5   C

6   Part A: 2.8 m;   Part B: 2,800 mm;   Part C: Longer by 0.8 m (80 cm).   7 $\frac{7}{4}$ or $1\frac{3}{4}$ 8   288

9   B   10   D   11 $5^2 = 25$ square units   12   B   13   (a) 9 and 1; (b) 5; (c) 4 terms   14   B

15   A   16 $3 + 2m$ 17   A   18 $p \leq 12$ and $w \leq 2,000$ 19   C   20   B   21   B   22   B

23   $(9, -5)$ 24   A   25   B   26   Yes, the data is roughly symmetric.   27   B   28   C   29   A

Find more at
ViewMath.com/AL-Grade6

 **30** *B*

💡 **Time to Learn!** 💡

*Review the explanations below, **especially for the questions you missed**.*

Understanding why each answer is correct builds stronger problem-solving skills.

***Tip:** Circle any questions you got wrong, then read their explanation carefully.*

📖 Practice Test 10 — Detailed Explanations

1. Apple has 3 parts, orange has 5 parts. Total parts = 8. Total ounces = $8 \times 4 = 32$.

2. Store A: $\$4.50 \div 10 = \$0.45$. Store B: $\$3.20 \div 8 = \$0.40$. Store B is cheaper per pencil.

3. $2 : 3$ multiplied by 3 gives $6 : 9$. The other pairs do not simplify to $2 : 3$.

4. The ratio is $2 : 3$. Choice A continues with $6 : 9 = 2 : 3$. Choice B has $6 : 8$ which is not $2 : 3$.

5. 48% is close to 50% = 100. Exact: $0.48 \times 200 = 96$.

6. Part A: $280 \div 100 = 2.8$ m. Part B: $280 \times 10 = 2{,}800$ mm. Part C: $2.8 - 2 = 0.8$ m longer.

7. $\dfrac{7}{10} \times \dfrac{5}{2} = \dfrac{35}{20} = \dfrac{7}{4} = 1\dfrac{3}{4}$.

8. $60 \div 21 = 2$ R18; bring down 4 → $184 \div 21 = 8$ R16; bring down 8 → $168 \div 21 = 8$. Answer: 288. Check: $288 \times 21 = 6{,}048$.

Find more at
ViewMath.com/AL-Grade6

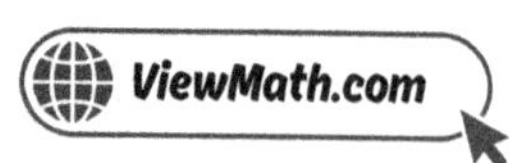

9. On a horizontal number line, negative numbers are always to the left of zero and positive numbers are to the right.

10. The absolute value of 0 is 0 because zero is 0 units from itself. Zero is the only number whose absolute value is 0.

11. The square has side length 5, so area $= 5^2 = 25$ square units.

12. Adding 12 to h gives a height that is 12 inches more than Hannah's.

13. Terms with y: $9y$ (coefficient 9) and y (coefficient 1). The constant is 5. There are 4 terms total.

14. $10 - 2(4 - 1) = 10 - 2(3) = 10 - 6 = 4$.

15. Distribute: $7k - 21 + 5$. Combine: $-21 + 5 = -16$. Result: $7k - 16$.

16. Flat fee: $3. Per-mile cost: $2m$. Total: $3 + 2m$.

17. Subtract 3.5: $x = 10 - 3.5 = 6.5$.

18. "Maximum 12 persons" means $p \leq 12$. "Maximum weight 2,000 lbs" means $w \leq 2,000$.

19. Open circle at 0 means 0 is not included. Shading right means greater than: $x > 0$.

20. $24 = \frac{1}{2} \times 8 \times h$, so $24 = 4h$, which gives $h = 6$ ft.

21. $A = b \times h = 9 \times 5 = 45$ cm^2.

Find more at
ViewMath.com/AL-Grade6

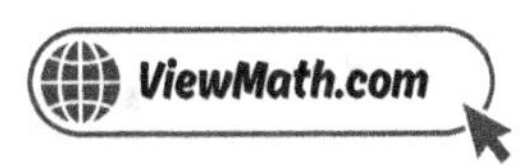

22 $V = 3 \times 2 \times 2 = 12 \ ft^3$.

23 The fourth vertex must share an x-value with $(9, 3)$ and a y-value with $(2, -5)$, giving $(9, -5)$.

24 Bottom rectangle: $8 \times 3 = 24$. Top-left rectangle: $4 \times (7 - 3) = 4 \times 4 = 16$. Total: $24 + 16 = 40$ square units.

25 Students live at different distances, so travel times vary. This is what makes it a statistical question.

26 The values are evenly spaced and balanced around the center (65). The left and right sides mirror each other.

27 The data is already in order. The middle (3rd) value out of 5 is 9.

28 3 hours: 5 dots, 4 hours: 3 dots, 5 hours: 2 dots. Total $= 5 + 3 + 2 = 10$ students.

29 Min $= 8$. Q1: median of $8, 12, 14 = 12$. Median $= 18$. Q3: median of $22, 26, 30 = 26$. Max $= 30$.

30 A good summary combines center ("typical value is about 20") and spread ("span 10 units").

Well done checking your answers!

Keep practicing to strengthen your skills.

Author's Final Note

I hope you enjoyed this book as much as I enjoyed writing it. Whether you are a student working through the material, a parent supporting your child's learning, or a teacher guiding your class, I have tried to make this book as clear and engaging as possible. I hope I have succeeded. If you have any suggestions for improvement, please let me know. I would love to hear from you.

The accuracy of calculations is very important to me. We have done our best, but I also expect that I have made some minor errors. Constant improvement is the name of the game. If you find any errors, please let me know. I will fix them in the next edition.

For students: Your learning journey does not end here. I have written a series of books to help you learn math. Make sure you browse through them. I especially recommend workbooks and practice tests to help you prepare for your exams.

For parents: Thank you for investing in your child's education. I encourage you to explore the companion resources available online to help support your child outside the classroom.

For teachers: Thank you for the invaluable work you do every day. I hope this book serves as a useful resource in your classroom. Feel free to reach out if you have suggestions or would like to discuss how best to use this book with your students.

I also enjoy reading your reviews. If you have a moment, please leave a review on where you found this book. It will help others find this book. If you have any questions or comments, please feel free to contact me at drNazari@ViewMath.com.

And one last thing: Remember to use online resources for additional help. I recommend using the resources on https://ViewMath.com You can find video lessons, practice problems, and more. You can also use the online companion for this book to track your progress and access additional resources.

Wishing all students the best in their studies, parents every success in supporting their children, and teachers continued inspiration in their classrooms!

Dr. A. Nazari

 **Great Job! Keep Learning with ViewMath!**

Keep up the great work! Visit **viewmath.com/AL-Grade6** for free lessons, quizzes, and more.

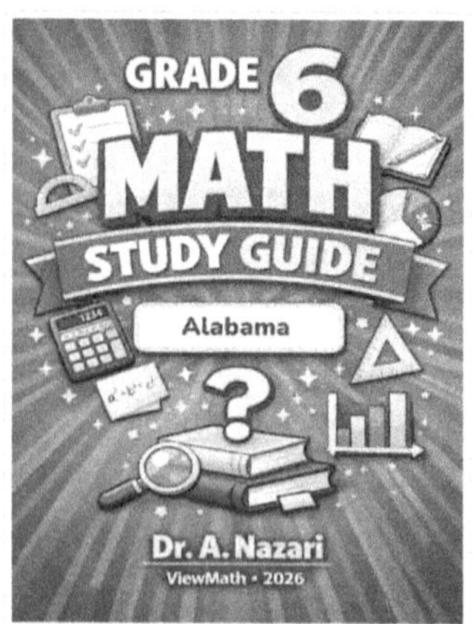

Study Guide

Workbook

Step-by-Step

3 Practice Tests

5 Practice Tests